# THE PHILOSOPHICAL SCIENTISTS

# David Foster

# The Philosophical Scientists

BARNES
&NOBLE
BOOKS
NEW YORK

This edition published by Barnes & Noble, Inc.,
by arrangement with C. Hurst & Co.

1993 Barnes & Noble Books

ISBN 0-88029-624-0

Printed and bound in the United States of America
M  14  13  12  11  10  9  8  7

# CONTENTS

# INTRODUCTION

This is an odd sort of book if only because it took 32 years (1952–84) to write. Between 1937 and 1950 I was a student of the ideas of Gurdjieff and Ouspensky, both of whom I knew, but by 1950 I decided that their ideas lacked something, namely a failure to take into account the developments in modern science. Accordingly I decided to go my own way. Why I then took 32 years to complete the book is explained by the need to wait for clarification of certain vital developments in science: notably the clarification of molecular biology in the 1950s and of the 'software' ideas from cybernetics in the 1960s and 1970s.

I started this book with the realisation that the developments in modern physics had opened new doors to philosophical thought. The centre for such ideas was Cambridge University and the ideas being put forward by Eddington, Jeans and Whitehead in the 1930s (what I call 'the 1930 Cambridge Club'), whose main tenet was

'The stuff of the world is mind-stuff.'

They had come to the view essentially because mathematics was more real than things. Substance did not exist. Mechanical world models were outdated. Space was curved.

The scientists had strayed into a new sort of territory and had stepped out of a physical universe of substance into a universe of mathematics. Furthermore the mathematics were very strange, being allied to concepts which were barely rational in the normal understanding of that word. On the whole, geometrical concepts were replacing arithmetic and algebra, but the geometry was non-Euclidean: an extreme example was Einstein's matter-tensor in which mass-energy could be replaced by fields of curved space.

No-one could find an adequate name for these new explorations, and we were left with the somewhat undescriptive 'modern physics'. Nor was it metaphysics since the new mathematics *worked* and therefore was science. So what was happening was that a world of substance was being replaced by a world of mathematics as the basic reality.

## Supernatural science

While these ideas were very suggestive that reality was of an abstract and even 'mental' character, they were essentially inferential and lacked the sort of numerical proof that 'two and two makes four'. But over the last thirty years in the field of molecular biology there has been the most astonishing development in the demonstration that all organic life is programmed by the DNA and that the DNA is a *coded statement* and the code is specific (as specific as an alphabetical statement) and transcends chemical science. So we now have to admit *supernatural science*, with a supernaturality capable of numerical proofs.

For example I have worked out two supernatural values:

1. The specificity of the haemoglobin protein is represented by the number $10^{650}$. What this means is that if haemoglobin evolved by chance there would only be one chance in $10^{650}$ of it actually occurring.

2. The specificity of the DNA of the T4 bacteriophage is represented by the number $10^{78,000}$ so that there is only one chance in $10^{78,000}$ of it actually occurring by random shufflings.

These figures have to be set against the fact that the universe is only $10^{18}$ seconds, old, and so there is no possibility whatsoever of life having evolved through Darwin's theory of natural selection operating on chance mutations. Furthermore, Darwin's main supporter T. H. Huxley had the theory of evolution completely wrong. In support of Darwin he stated that if six monkeys strummed at random on typewriters for 'millions of millions' of years they would type all the books in the British Museum.

If we assume that 'millions of millions' of years is the life of the universe at 10 billion years, then a modern computer will tell us that those monkeys would only type out *one half-line* of sense and with the choice of matching any line in all those 700,000 books in the British Museum.

It is a debacle, a debacle of 'scientism' which made statements without doing the corresponding calculations. Nor can they be excused because they did not have modern computers, since logarithms had been available for 200 years before 1860 and the calculations could have been done. One can only hope that Bishop Wilberforce will now sleep a little more comfortably in his grave.

I did not set out in this book to destroy Darwinism. I set out to explore the plausibility of the 1930 philosophical scientists who were declaring that 'the stuff of the world is mind-stuff', and by the end of the book I find the case proven from the new knowledge from molecular biology (in major fashion) and from cybernetics (in minor fashion).

But once one discovers the supernatural in science, the mind is opened to further possibilities. I had stumbled on the importance of *specificity* in biology by accident: I had read a great deal, but I became increasingly impressed by H. F. Judson's *The Eighth Day of Creation*, published in 1979, which had been recommended to me as compulsory reading by Professor Frederick Sanger of Cambridge, double Nobel Laureate and the first person to unravel the *specificity* of a protein (bovine insulin). Judson's book is a work of art and in surveying the historical development of molecular biology he initiates a minor theme as to 'specificity'. But as the book develops this becomes SPECIFICITY until by the end we are convinced of SPECIFICITY!!!!!!!!!!!! It is *new*. It deals body-blows to Darwin and to the Second Law of Thermodynamics.

## *The creation of the Universe*

But then as a result of my reading I came across a quite different reference (in Condon and Odishaw's *Handbook of Physics*) and this was Einstein's matter-tensor $T_{ik}$. This equates mass-energy ('matter') with a purely geometric equation or 'form'. Put shortly, if God could 'shape the void', then — hey presto — one would have the actual physical universe without any need for some sort of original Big Bang energy. The universe would simply be 'thought up'. I would emphasise that I regard all such as speculative metaphysics and not the main feature of this book, which is *supernatural science* based essentially on modern molecular biology. But given the *fact* of *supernatural science*, then the horizons of plausibility expand into the bizarre, and non-Euclidean geometry is bizarre. It is the great hope for the future of society which has been deceived by scientism into agnosticism, atheism and 'God is dead'.

We are on the verge of new realisations.

*The Glade, Ascot, England*                                      D. F.

# PART ONE
# THE 1930 CAMBRIDGE CLUB

Before the Second World War of 1939–45 there was developing at Cambridge a new philosophy from physics, perhaps mainly identified with Sir Arthur Eddington. It was based on the revolutionary new ideas in physics coming from Einstein, Planck and Heisenberg and which compelled Eddington to state: 'Religion first became possible for a reasonable man of science in the year 1927.' The thesis was 'The stuff of the world is mind-stuff.' This new-look remains valid, and Sir Edmund Whittaker writing in 1955 stated that it is still the most lucid philosophy from science that we have. Unfortunately the outbreak of the 1939–45 war diverted scientific thought into military endeavour, including the atom bomb, and this philosophy was sunk without trace.

But in retrospect the implications are far-ranging.

# 1

## THE 1930 CAMBRIDGE CLUB

The great developments of physical science in this century took place over the years 1900 to 1927. The developments were essentially in the field of mathematical physics and came from a German school (including Swiss and Danes) associated with the names of Planck, Einstein, Bohr, Born, Sommerfeld, Heisenberg, Schrödinger and Pauli. The main developments were relativity theory, quantum physics, wave mechanics and the uncertainty principle. What they did was to destroy the mechanical picture of nature, including strict causality, which had been inherited from Newton.

The impact on the human mind was traumatic. When we try to understand something complex, we like to use our visual imagination and make a map or model of the problem. But the New Physics was purely mathematical, and all models were barred unless one can treat a mathematical equation as a model and very few people can do that.

The German school which had originated the New Physics did not particularly philosophise about its findings. But in Britain over the same period there was a group of four people who, in their different ways, had quite a bit to say. They were Sir Arthur Eddington, Sir James Jeans, Bertrand Russell and A. N. Whitehead.

Cambridge University has made outstanding contributions — typically the discovery of the electron and the splitting of the nucleus which eventually led to atomic energy. But it was also noteworthy for linguistic philosophy and the analysis of the meaning of statements. These four men did not comprise a school since they did not appear to have a common objective, but they knew each other and influenced each other and so might be called a club. I call it 'the 1930 Cambridge Club' because it was responsible for philosophical ideas relevant to the New Physics and because it peaked about the year 1930. The outstanding statement was by Sir Arthur Eddington in his book *The Nature of the Physical*

*World.*\* This was published in 1928, and in his Introductory Note
to the 1955 edition† Sir Edmund Whittaker, FRS, wrote:

> From the time of its first appearance *The Nature of The Physical
> World* has been accepted as the most profound, and at the same
> time the most lucid, treatment of the philosophy of the new
> physics. It may seem surprising that the developments in
> science since its publication in 1928 have not brought such
> changes in our understanding of nature as to make the book
> more or less out of date: but the fact is that while in some
> directions of research there has been great progress — e.g. the
> recognition of the neutron, the positron, and many kinds of
> meson — these discoveries have had little or no influence on
> philosophy; the great change in philosophical outlook followed
> from the scientific work of the years 1900–27, and was asso-
> ciated with the introduction of special relativity, general rela-
> tivity, quantum theory and the uncertainty principle. At the
> end of this period Eddington, in his Gifford Lectures at
> Edinburgh, created a radically new philosophy of science which
> has been dominant ever since, and it is this that is presented in
> *The Nature of The Physical World.*

But in the 1980s it would not be true to say that Eddington's
views dominate philosophy related to science. Not that they have
been replaced but rather that there is no longer any significant philosophy in the world of science. His philosophy did not die of old
age, nor was it faulted, but it seems that the Second World War
switched science on to different tracks of a military nature. Expenditure in science is still dominated by military requirements.

Since *The Nature of the Physical World* there have been only two
other books, both from the same author, to compete with it as a
broad sweep of philosophy applied to science. These are Erwin
Schrödinger's *What is Life?* and *Mind and Matter*, published res-
pectively in 1944 and in 1958. Schrödinger's writings are an

\*Cambridge University Press, 1935.
†Everyman's Library, J. M. Dent & Sons.

attempt to embrace physics with biology: we shall consider them later in this book. It is my impression that Schrödinger took over Eddington's mantle as the leading philosophical scientist of the following generation, and his views are deeply respected by both physicists and biologists.

## The four revolutionary ideas of modern physics

The four new ideas of modern physics mentioned above by Sir Edmund Whittaker were:

### SPECIAL RELATIVITY (EINSTEIN)

It had long been assumed that light was a vibration in some sort of background ether, rather like sea-waves are waves in the water of the ocean. But laboratory measurements by Michelson-Morley in 1887 showed conclusively that although light waves certainly existed they were not a vibration of an ether: ether did not exist and light waves were only vibrations of themselves. Einstein saw that the inference from this was that the concept of 'absolute position' or 'absolute clock time' was meaningless and that the only constancy was that of the velocity of light which never changed no matter who observed it or what was the speed of either the light source or the observer. I am not convinced that anyone (including myself) has ever understood Einstein's theory of Special Relativity, for the fact is that it does not lend itself to a mechanical model or analogue, and what matter are the mathematical equations which Einstein developed relating space, time and the velocity of light and *which work*. Of such implications the main one is that any object moving with the speed of light would have infinite energy and weight. This has been proved in the laboratory.

### GENERAL RELATIVITY (EINSTEIN)

The General Theory replaces the 'action at a distance' as to the gravitational attraction between masses by the idea that masses distort space-time so that it curves in such a way that the track of a particle will be exactly the same as though it were attracted by gravitational action at a distance. Thus gravitational forces are replaced by the change of the nature of space-time in the presence of masses. Einstein forecast that this would mean that light would be bent when it came near to masses, and this was demonstrated many years later under solar eclipse conditions in that the light from a star just glancing the Sun on its way to Earth was detected as being bent, much as the theory predicted. The main upshot has been a number of mathematical equations *which work*.

Note that in neither of Einstein's theories can one make a model of the physical events (how could one model curved space?), but in the upshot there are useful mathematical equations.

## QUANTUM THEORY (PLANCK)

Experiments in the laboratory showed that radiation such as heat or light was not continuous. Planck showed that radiation is emitted in specific bursts of an exact and unvarying magnitude rather like bullets being shot from a gun. As applied to the theory of the atom and electronic orbits, the same specificity was to be described by integral numbers (quantum numbers), not unlike the way in which musical scales are formed from integral numbers so that the octave is quantified by the integral ratio 2/1 and the major fifth by the integral ratio 3/2. Here again we cannot make a model, but the mathematics *work*.

## THE PRINCIPLE OF UNCERTAINTY (HEISENBERG)

It used to be assumed that micro-events were of a causal continuous nature like a train pushing coal trucks. Heisenberg showed that if an event had two measurement parameters (such as position and momentum) one could put an exact value on one only at the expense of vagueness about the other and vice versa. The analogue would be that of a camera which can only focus clearly on one plane of vision by being out of focus on another plane closer or further away. The significance of this was that it was impossible to calculate precise cause-and-effect between two micro-events because there was an element of uncertainty or out-of-focus in the calculations. The out-of-focus had to be replaced by probability mathematics *which worked*.

So in considering the above four main discoveries which have revolutionised modern physics, in each case the basic principle is obscure but the mathematics *work*. It is probably true to say that all the above four notable discoveries were the result of inspired guesswork. One even suspects the mathematics came before the explanations.

As Sir Edmund Whittaker noted, it is these four discoveries which led to the philosophising of 'the 1930 Cambridge Club'. This is what we consider next.

# 2

## EDDINGTON AND 'THE STUFF OF THE WORLD IS MIND-STUFF'

The human being only feels at home in an environment where there is solid ground underfoot. The human being also likes to be surrounded by walls and a roof which are solid compared with the wind and rain which might otherwise be disturbing. Accordingly our houses are made from solid materials such as wood, brick or glass. The same preference for solidity extends to the larger environments, to roads and walkways and to shops and offices and workshops and wherever our bodies have a position in space.

This desirable solidity of our familiar environment is not quite the same as 'substantiality'. Water is substantial but most of us would not wish to try standing on it. But solidity is relative to solid-sensing and is not an absolute property. So the fly that one sees crawling up the wall and then across the ceiling is not afraid to fall off since for a fly the air of the room is solid to its wings. We may also suspect that water is solid for a fish.

So just what is solidity? It would seem to be that property of the environment which is sufficiently strong, durable and stable to be able to support our physical bodies in a gravitational field. According to this definition air, water and the ground are 'solid' according to whether we are considering the environment of a fly, a fish or a man.

### Solidity and substance

Solidity is a relative property, but of what? We may conceive the notion of Substance as that 'thing' which can have properties such as solidity. Sir Arthur Eddington stated:*

*All quotations in this chapter are from *The Nature of the Physical World* by Sir Arthur Eddington (Cambridge University Press, 1935).

One of our ancestors, taking arboreal exercise in the forest, failed to reach the bough intended and his hand closed on nothingness. The accident might well occasion philosophical reflections on the distinctions of substance and void, to say nothing of the phenomenon of gravity. However that may be, his descendants down to this day have come to be endowed with an immense respect for substance arising we know not how or why. So far as familiar experience is concerned, substance occupies the centre of the stage, rigged out with the attributes of form, colour, hardness, etc., which appeal to our several senses.

So strongly has substance held the place of leading actor on the stage of experience that in common usage *concrete* and *real* are almost synonymous. Ask any man who is not a philosopher or a mystic to name something real; he is almost sure to choose a concrete thing. Put the question to him whether Time is real; he will probably decide with some hesitation that it must be classed as real, but he has an inner feeling that the question is in some way inappropriate and that he is being cross-examined unfairly.

Our preference for substance is due to the protection it affords us against the fearful force of gravity. We have only to go near to the edge of a steep cliff to feel our instincts quake at the fatal consequences of lack of solid substance under our feet if we step over the edge. Thus our respect for solid substance is no mere mental notion but is the instinctive understanding of what is good for us, namely to have solid ground under our feet to support us against the free-falling of gravity.

## The Concept of Matter

Given the assumption of solid substance, we then make a further assumption that it could be divided into smaller and smaller parts all of the same nature; this aspect we call Matter. Thus solid substance is supposed to be made of an aggregate of matter just as the seashore is made of grains of sand. Now this is true only down to a certain level of smallness, represented by crystals and molecules. But when one looks for the fine-structure of molecules as to their atoms, one enters a region dominated by void or emptiness: and ultimate fine-structure is not structure at all but consists of electrical and gravitational fields cavorting in the void. At the fine-structure level, and just when one might expect to find ultimate particles of matter, the matter has vanished!

So matter is not a thing, but it is a *state of organisation* which at a certain level of integrated organisation reveals properties such as solidity according to the experience of appropriate creatures. It is all very subjective and relative, and we have noted that 'solid' ground, water and air are equally 'solid substance' according to whether one is a human being, a fish or a fly.

## Actuality

We have the following facts:

For a human being, terra firma is 'solid substance'.

For a fish, water is a relatively 'solid substance'.

For a fly, air is a relatively 'solid substance'.

The common factor in the above statements is that there appears to be an experience which is describable, but which is based on different physical facts for different creatures. This common factor can be called *Actuality*, 'what works'. Actuality is a very important category, and the human relationship with the world is dominated by it; it could even be described as relative reality, that reality which is true for a given sort of creature.

Nevertheless, there might seem to be just a little of common physical fact to the three aspects of solidity and especially as to the dependence on the density of matter. But if we asked a physicist to put density figures to our actualities he would reply:

*Human-solidity* as to (say) a brick wall implies a mass density of about 4 grams per cubic centimetre.

*Fly-solidity* as to air implies a mass density of .0013 grams per cubic centimetre.

Thus if the physicist tries to quantify solid actuality, he comes up with two figures which are a ratio of 3,000 to 1 apart. Clearly the physicist cannot help us much when we come to consider Actuality. As Eddington stated:

> In the scientific world the concept of substance is wholly lacking, and that which most nearly replaces it, electric charge, is not exalted as star-performer over the other entities of physics. For this reason the scientific world often shocks us by its appearance of unreality. It offers nothing to satisfy our demand for the concrete . . . in leading us away from the concrete, science is reminding us that our contact with the real is more varied than was apparent to the ape-mind, to whom the bough which supported him typified the beginning and the end of reality. The modern scientific theories have broken away from the common standpoint which identifies the real with the concrete . . . It would not be fair, being given an inch to take an ell, and say that having gone so far physics may as well admit at once that reality is spiritual. We must go more warily.

## The world of physics — a closed system

The world of physics is a world based on laboratory measurements which result in pointer readings and numerical values. Within this system there have been discovered certain invariable numerical constants which act as its anchor-points. These are:

The velocity of light
Avogadro's number
Planck's constant

The mass of the neutron
The charge on the electron.

Given these basic units, then, their interplay can be expressed in mathematical equations to which numbers can be assigned. Because of the anchorage by the constants, the ultimate outcome is a set of Laws of Nature but, unlike Actuality, these are specific and fixed, and form a closed system. As Eddington stated:

> In science we study the linkage of pointer readings with pointer readings. The terms link together in endless cycle with the same inscrutable nature running through the whole.

But is such a world close to what might be called Reality and our familiar world of Actuality? It is so only in one respect, in that the world of physics has arisen out of human thought and can be manipulated by human thought, and so it contacts the world of Actuality *in the human mind.*

## Eddington's philosophical leap

Although Eddington agreed that there were two sorts of world, that of familiar Actuality and that of physics, he was unhappy that they had little connection except through the human mind and in that the same human mind which can appreciate the solidity of a brick wall is the same human mind which can appreciate that $E = mc^2$. The common link was *mind* and it was this fact which caused him to make his famous statement 'The stuff of the world is mind-stuff.'

Thus physics could be conceived as a particular form of Actuality. While retaining its pointer readings, constants and a fixed set of laws, all these could be appreciated within the human mind and thus form a special part of the human mind. He qualifies this:

> The mind-stuff of the world is, of course, something more general than our individual conscious minds; but we may think

of its nature as not altogether foreign to the feelings in our consciousness.

It seems he was also influenced towards his views by the evidence that mind-stuff can be unconscious or sub-conscious as in our temporarily forgotten memories and in dreams. This showed there were mind-stuff attributes below the level of normal human conscious experience, and of course psychologists such as Freud and Jung would have supported this point. But I think we have to leave Eddington's views as a broad intuition of a great scientist who was not prepared to divide the world into two as between familiar Actuality and the world of physics 'out there'.

For Eddington the fact that his desk on which he wrote was obviously solid, whereas his physics told him it was mainly void in which electrons and nucleons cavorted, could only be reconciled by considering them both as aspects of an Actuality which was based in the mind.

# 3

## JEANS AND THE UNIVERSE AS 'THE THOUGHT OF A MATHEMATICAL THINKER'

It was Sir James Jeans who supplied a complementary argument — from that realm of 'the cyclical metrical system' of physics which Eddington considered very limited and circumscribing for actuality. Indeed Eddington's views could be described as 'the argument from actuality', from human conscious and sub-conscious experience *now*. It is true that Eddington demonstrated the weakness of substance in science, but he retained the validity of subjective experience. Let us begin with a main statement by Sir James Jeans:*

> The essential fact is that all the pictures which science now draws of nature, and which alone seem capable of according with observational fact, are mathematical pictures. Most scientists would agree that they are nothing more than pictures — fictions if you like, if by fiction you mean that science is not yet in contact with ultimate reality. Many would hold that the outstanding achievement of twentieth-century physics is not the theory of relativity with its welding together of space and time, or the theory of quanta with its present apparent negation of the laws of causation, or the dissection of the atom with the resultant discovery that things are not what they seem; it is the general recognition that we are not yet in contact with ultimate reality.

Jeans explores this point by analogy with the people in Plato's Cave who, with their backs to the light, could only see shadows thrown on the wall of the cave by entities which moved along behind them. Thus the concept was that science and physics constitute the study of shadows of reality. But something is to be dis-

---

*All quotations in this chapter are from *The Mysterious Universe* by Sir James Jeans (Cambridge University Press, 1930).

covered from the study of shadows. The shadow of a falling body is something like the actual falling body. The shadow of the Knight's Move is something like an actual Knight's Move in the game of chess. But what is extraordinary is that the shadows are mathematical.

> Nature seems very conversant with the rules of pure mathematics, as our mathematicians have formulated them in their studies, out of their own inner consciousness and without drawing to any appreciable extent on their experience of the outer world.

Having established a general analogy of scientific reality, being alike to the shadows thrown on the wall of Plato's Cave and therefore almost certainly being different from the facts of the objects throwing the shadows, Jeans takes a quantum leap by pointing out that when the shadows are of a mathematical nature then they are familiar to the cave-dwellers. Thus if the shadow was that of a motor-car the cave-dwellers would be perplexed. But if the shadow was the mathematical formula $E = mc^2$ then they can declare that they know all about that. The paradox would be explained if human beings were born mathematical games-players and the external world was a theatre of mathematical games-playing. That mathematics might be more fundamental than 'the thing in itself' had also been stated by Einstein in a conversation with Heisenberg: 'It is the theory which decides what we can observe.'

We tend to think of mathematics as something which has been derived from experimentation, but Jeans points out that pure mathematics is a product of pure human thought. For example, the Euclidean proposition that the angles of a triangle make up two right angles cannot be exactly observed in nature but it can be derived through logic. What is novel is not so much that the basis of physical reality is mathematical, for most scientists have little choice but to adhere to such a view, but rather that the most abstract thinking of human beings is in tune with truths from physical reality represented by mathematics rather than things.

This agreed with Eddington (previous chapter) who suggested that ultimate sub-consciousness attributable to inanimate matter would be broadly of the same nature as human psychic activity. But Jeans was being more specific in claiming that its form would be mathematical:

> A scientific study of the action of the universe has suggested a conclusion which may be summed up, though very crudely and quite inadequately . . . that the universe appears to have been designed by a pure mathematician.

It could be objected that because human beings can be mathematicians they are projecting this faculty on the world, just as a person wearing blue glasses would see a blue world. But Jeans refutes this:

> A moment's reflection will show that this can hardly be the whole story. Our remote ancestors tried to interpret nature in terms of anthropomorphic concepts of their own creation and failed. The efforts of our nearer ancestors to interpret nature on engineering lines proved equally inadequate. On the other hand our efforts to interpret nature in terms of the concepts of pure mathematics have, so far, proved brilliantly successful . . . from the intrinsic evidence of his creation, the Great Architect of the Universe now begins to appear as a pure mathematician.

## The universe as mathematical waves

Jeans had a very simple concept or model of physics. He points out that in the nineteenth century the anchor-points had been three laws of conservation — conservation of mass, of energy and of matter — but by the early part of this century these three had been replaced by a single conservation law of mass-energy with mass and energy being inter-convertible. But this simplification led to a further simplification since all the mathematics was pointing to a wave-like reality and this had been driven home by Schrödinger in wave-mechanics; he had shown that 'solid matter' had waves, and the practical proof of this had been the electron microscope where electrons behaved like light waves.

But what Jeans suggested was that all radiation such as light was linear waves, waves travelling in straight lines, while all 'particles' were rotating or spinning waves. Thus we could anticipate atomic power if the spinning waves could be set free and become the linear waves of radiation. Physical developments since the time of Jeans have rather supported this analysis. It is an attractive theory, since it is easier to conceive of the energy of mass being 'actual energy' trapped in spin like the motion of a flywheel rather than some innate property of mass of an esoteric nature. However, the speculation has been neither proved nor disproved.

## The universe as thought

Jeans' final words on his beliefs are:

> If all this is so, then the universe can be best pictured, although still very imperfectly and inadequately, as consisting of pure thought, the thought of what, for want of a wider word, we must describe as a mathematical thinker. If the universe is a universe of thought, then its creation must have been an act of thought. Modern scientific theory compels us to think of the creator as working outside time and space which are part of his creation, just as an artist is outside his canvas. The old dualism of mind and matter seems likely to disappear, not through matter becoming in any way more shadowy or insubstantial than heretofore, or through mind becoming resolved into a function of the working of matter, but through substantial matter resolving itself into a creation and manifestation of mind.

Thus Jeans joins Eddington as to 'the stuff of the world is mind-stuff', but emphasising the mathematical nature of the mind-stuff.

# 4

## RUSSELL: 'MATHEMATICS AND LOGIC ARE IDENTICAL'

The third participant in the 1930 Cambridge Club was Bertrand Russell. In his *Principles of Mathematics* (1903) he wrote:

> The fundamental thesis of the following pages, that mathematics and logic are identical, is one which I have never seen any reason to modify.

This statement is of great interest and significance related to the slightly different views just described from Eddington and Jeans. Eddington took the view that 'the stuff of the world is mind-stuff', and it appears he had in mind that it was the sort of logical-linguistic mind-stuff which we human beings experience as thinking. Jeans took the view that the mind-stuff was more of a mathematical nature. But if logic and mathematics are the same thing, then Eddington's and Jeans's views are identical.

Russell's great books on the subject were lengthy, and as well as the one just referred to he also wrote (with A. N. Whitehead) the three-volume *Principia Mathematica* (1905). In the latter book great use was made of symbolic logic and the special symbols which that employs.

But while credit must be given to Russell (and to Whitehead to some degree), the fact is that they were pre-empted by George Boole in about 1840 and post-empted by the growth of computer technology in the present era. For this reason I shall not give the Russell proofs as to the identity of logic and mathematics (which are very lengthy), but will give a brief version of the work of Boole and modern computery.

## Boolean Algebra

George Boole had realised that if one had data which was True or False and if one then processed this data with thinking which could be True or False, then one could write a very simple sort of equivalent algebra (Boolean Algebra):

Let True take the cipher 1.
Let False take the cipher 0.

Then one can construct a Truth Table:

|  | *Algebra* |
|---|---|
| True data with True thinking = True | $1 \times 1 = 1$ |
| True data with False thinking = False | $1 \times 0 = 0$ |
| False data with True thinking = False | $0 \times 1 = 0$ |
| False data with False thinking = False | $0 \times 0 = 0$ |

A little scrutiny shows that the statements given on the left in logical terms are the same as the statements on the right in mathematical terms. From this very simple logical-mathematical equivalence the whole of modern electronic computing has developed.

## Modern electronic computers

A modern electronic computer simply consists of an array of electronic switches. If a switch is closed it stands for the cipher 1 and if a switch is open it stands for the cipher 0. By arranging switches as ciphers one can create sequential codes of any length which can represent either alphabetical words or numbers. The same computer can be programmed to play a logical game such as a game of chess, or it can be programmed to find all the prime numbers up to one million or more. To my mind it is the practical fact of electronic computers, which can do logical-mathematical tricks, which is the central proof that logic and mathematics are the identical subject. The stark proof in elegant simplicity is the Boolean Truth Table just given.

Thus Eddington and Jeans were saying the same thing in deciding that 'the stuff of the world is mind-stuff.'

# 5

## WHITEHEAD AND ORGANIC MECHANISM

Although A. N. Whitehead worked with Russell on symbolic logic, he also made a major contribution of his own through his ideas about organic mechanism, as described in his book *Science and the Modern World** from which the quotations in this chapter are taken. The important contribution was to take the ideas of Eddington and Jeans, which were mainly related to inorganic physics, and create a bridge to living creatures. This, as it were, put an added dimension into the total situation being developed.

Whitehead's most general thesis is:

> The field is now open for the introduction of some new doctrine of organism which may take the place of the materialism with which, since the seventeenth century, science has saddled philosophy. Science is taking on a new aspect which is neither purely physical nor purely biological. It is becoming the study of organisms. Biology is the study of the larger organisms: whereas physics is the study of the smaller organisms. This is the theory of organic mechanism.

Whitehead virtually abolishes matter (as did Eddington and Jeans, as we have seen) and replaces it by organisation and organism. His sort of argument is: if I kick the leg of my table, am I kicking matter or a table? For practical purposes I am kicking the table, and although one could theorise that the table is made of matter, there is no known way of describing it or analysing the wood of the table down the ladder of cellulose, molecules, atoms and electrons where one can state 'This is matter'. In fact there is simply a ladder of *organisation* which ultimately is 'a leg of a table' as human actuality.

*Cambridge University Press, 1926.

## The nature of organism

Whitehead derives his idea of organism from logical premises integrating several different points of view about the same reality and with the notion that the total point of view is more than the sum of the parts. One might cite the example of the blind men in India who were examining an elephant:

One felt the leg and said 'It is a tree'.
Another felt the ear and said 'It is a fan'.
A further felt the trunk and said 'It is a water pipe'.
One more felt the tail and said 'It is a rope'.

The organic step is taken when someone puts all this together and suggests 'Actually, it is an elephant'.


## The general schematic of organism

A general schematic of organism is as shown in Fig. 5.1:

1. There are a set of peripheral observers such as A, B, C, D and E.

2. Each of such observers will be conditioned by relationships with all the others as shown by the ten internal lines of connection or network.

3. The integrative organic meaning in such a system is the totality of the network as observed by a notional central observer. This is not shown in Fig. 5.1 since it is co-incident with the network pattern considered as a whole, i.e. the central observer must be 'everywhere' to take into account all peripheral points of view.

The above illustration is somewhat simpler than the analysis of organism made by Whitehead. But it is not different in principle, since Whitehead assumes that every entity has a point of view conditioned by its environment and with the totality of points of view creating the organic.

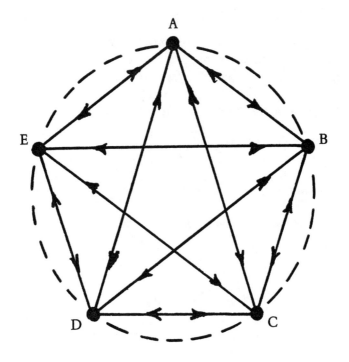

Fig. 5.1. Schematic of organism.

## The bridge between the inorganic and organic

Whitehead examines the role of an electron in the human body as to what laws it obeys and first quoting a line from a poem by Tennyson:

> *'The stars'*, she whispers, *'blindly run.'* That line starkly states the whole philosophic problem implicit in the poem. Each molecule blindly runs. The human body is a collection of molecules. Therefore the human body blindly runs, and therefore there can be no individual responsibility for the actions of the body. If you once accept that the molecule is definitely determined to be what it is, independently of any determination by reason of the

total organism of the body, and if you further admit that the blind run is settled by the general mechanical laws, there can be no escape from this conclusion. But mental experiences are derivative from the actions of the body, including of course its internal behaviour. Accordingly the sole function of the mind is to have at least some of its experiences settled for it, and to add such others as may be open to it independently of the body's motions, internal and external.

There are then two possible theories as to mind. You can either deny that it can supply for itself any experience other than those provided by the body, or you can admit them . . . if the volition affects the state of the body, then the molecules in the body do *not* blindly run. . . .

The doctrine I am maintaining is that the whole concept of materialism only applies to very abstract entities, the products of logical discernment. The concrete enduring entities are organisms, so that the plan of the *whole* influences the very characters of the various subordinate organisms which enter into it. In the case of an animal, the mental states enter into the plan of the total organism and thus modify the plan of the successive subordinate organisms until the ultimate smallest organisms, such as electrons, are reached. Thus an electron within a living body is different from an electron outside it, by reason of the plan of the body.

What Whitehead suggests is not that an electron within the human body does not blindly run but rather that it blindly runs *differently* due to being within the body. The analogue of this might be a railway train which 'blindly runs', but just where it runs will depend upon the rail tracks of the railway system involved. So Whitehead is not trying to destroy the mechanical behaviour of electrons and molecules but is stating that their actual behaviour will depend upon the system within which they are operating.

*The hierarchy of lawful organic events*

Let us start at the bottom of the hierarchical system and first state that we know that an electron is governed by the laws of the atom and that atoms are governed by the laws of molecules. It would be no difference of principle to assume that molecules are governed by the large organic molecules of organic cells, the cells by the organs of the body which they constitute, and the organs by the whole body. In this fashion a volition from the whole body would travel down the same hierarchy so that it affected how the electrons 'blindly run'.

In considering the views of Eddington and Jeans that 'the stuff of the world is logical mind-stuff', we now find that Whitehead has added a new dimension to the concept of mind-stuff in that it is arranged in organic hierarchies. Common sense tells us this must be true, otherwise a human being comprising trillions of atoms would be a chaos without simplicity of experience. The human body comprises at least the following well-defined levels:

Electrons
Atoms
Molecules
Organic molecules
Cells
Tissues
Organs
Whole body.

The above could not possibly work unless each level had both its own laws and interaction laws so that it obeyed the volition and purposes of the level above. So the electron may blindly run but on the railway lines determined by my conscious intents.

*Erwin Schrödinger*

Erwin Schrödinger in *What is Life?* wrote something very similar to the last statement above:

1. My body functions as a pure mechanism according to the Laws of Nature.

2. Yet I know, by incontrovertible direct experience, that I am directing its motions. . . . The only possible inference from these two facts is, I think, that I — I in the widest meaning of the word, that is to say, every conscious mind that has ever said or felt 'I' — am the person, if any, who controls the 'motion of the atoms' according to the Laws of Nature.

# 6

## CONCLUSIONS FROM THE 1930 CAMBRIDGE CLUB

There are two distinct developments in philosophy coming from the 1930 Cambridge Club:

1. From Eddington and Jeans, reconciled by Russell, that 'the stuff of the world is mind-stuff'.

2. From Whitehead, that the basic nature of the universe is a hierarchical structure of organisms.

These ideas are not obviously similar, but that is the next point to explore.

### Eddington and Jeans

Eddington's argument centres on his need to find a reconciliation as between familiar Actuality (tables, chairs etc.) with his world of scientific Reality which is typified by the pointer-readings of atomic physics. The reconciliation he chooses is that both worlds belong to the common world of human percepts and concepts, i.e. both worlds are 'in here' in our heads. Thus there is no fundamental difference as between my idea of a table and my idea of an atom. So in considering tables and atoms Eddington is not differentiating kind but only scale and form. At the same time he is not totally subjective and suggests that the sort of consciousness which we consider typically human may extend to everything else in the universe. Thus he comes to a sort of objective-subjectivity in which mind is everywhere, and so to his statement that 'the stuff of the world is mind-stuff'.

By contrast Jeans' views are much more definite. His grounds are:

1. The reality of science is mathematical.

2. Mathematics must be admitted as the real 'things in themselves'.

So Jeans tightens up Eddington's rather vague mind-stuff proposition by suggesting that reality is centred around *symbolic data* as a general interpretation of what we mean by mathematics. This implies that the mind-stuff has specific forms or what we call symbols, including numbers and letters and their complexification.

## *Marks, symbols and language*

Let us consider in the most basic fashion how meaning can arise from chaos. Suppose we have a random set of dots as in Fig. 6.1. We can take these dots and arrange them in a pattern such as Fig. 6.2, which we recognise as the letter A. Alternatively we could take the random dots and arrange them in the pattern of Fig. 6.3 to form the letter B. If we make twenty-six alternative arrangements we can have the whole alphabet:

ABCDEFGHIJKLMNOPQRSTUVWXYZ

But if we take a choice of such letters (as in the game 'Scrabble') we could arrange them as to

THE CAT IS BLACK, or
I THINK IT WILL RAIN,

but now we are in the realm of intelligible meanings.

But what have we done in going from a set of random dots to clear literate meanings? We have proceeded through three stages:

1. Disorder of meaningless random dots, 'things'

2. *Arrangement* of the dots into letters

3. *Arrangement* of the letters into statements.

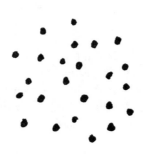

Fig. 6.1. Random dots.

Fig. 6.2. Dots arranged as letter 'A'.

Fig. 6.3. Dots arranged as letter 'B'.

So all we have done is to add *arrangement* to chaos. When we arrange a set of dots (Fig. 6.1) to be the letter A (Fig. 6.2), we have added something which is *more* than the sum of the parts. The simple sum of the parts is just the *number* of dots. But in arranging them as the letter A we have added a modicum of meaning and increased *potentiality* for meaning. When the letter A takes its place in the statement THE CAT IS BLACK, then the arrangement has added further meaning which is more than the sum of the parts. In this fashion we see that the world is a structure of *progressive arrangements* of a hierarchical nature, and such is essentially organisation or *organism*. This means that the letter A is not a 'thing', since we have transcended those random dots which are the basic 'thing' and have added a modicum of symbolic meaning. Number has transformed to literate meaning, and this becomes even more clear when we elaborate the letters to read THE CAT IS BLACK.

We have exactly the same state of affairs in considering living organic creatures such as an actual black cat. The black cat is not a 'thing', it is an arrangement of those random dots we call atoms. The fact that we use the words BLACK CAT is simply a naming process for a particular arrangement of atomic dots. It is symbolism.

We thus have an exact parallelism:

1. An organic creature is an arrangement of atomic dots.

2. A meaningful human statement is an arrangement of mark-dots.

## Eddington-Jeans and Whitehead

We see that Eddington-Jeans and Whitehead were coming to the same conclusion via different approaches. Eddington-Jeans were rather taking organism for granted and concentrating on the basic assumption of cosmic Mind with (Jeans) a bias towards mathematical Mind. By contrast Whitehead was tackling a more fundamental analysis, that of the nature of organism which could create Mind or even *be* Mind. So, as it were, Whitehead was working one step more fundamental and below Eddington-Jeans, but by working so he was underpinning the Eddington-Jeans outlook. But would Whitehead have agreed that 'the stuff of the world is mind-stuff?' We go round about a little to answer that question.

## The logical mathematics of arrangement

Whitehead's model of organism would be the diagram in Fig. 5.1 (page 22), in which a number of distinct entities had mutual relationships and were conditioned by each other and with the result that 'the whole is more than the sum of the parts'. Such a diagram is the same as that which operates for the calculation of the permutation of alternative arrangements of entities. In this we note five letters of the alphabet in a circle and ask how many arrangements of this there can be if we can move each letter into any place. The answer is $5 \times 4 \times 3 \times 2 \times 1 = 120$ arrangements ('permutations').

If we increase the number of entities, then the following table holds:

| Number of entities (such as letters of the alphabet) | Number of permutations ('words') |
|---|---|
| 1 | 1 |
| 2 | 2 |
| 3 | 6 |
| 4 | 24 |
| 5 | 120 |
| 10 | 3,628,800 |
| 20 | 2,432,901,600,000,000,000 |

So we see that the number of words, possibly meaningful, which can be created as permutations of a few letters is astronomical. All that this demonstrates is:

The ability to differentiate as between 'this and that' is virtually infinite provided we adopt the mathematics of the permutation of different entities (such as the letters of the alphabet).

Later in this book we shall see that this is just what Nature does in the field of molecular biology.

## Would Whitehead have agreed that 'The Stuff of the World is Mind-Stuff'?

We come back to the point whether Whitehead could have agreed with Eddington and Jeans that 'the stuff of the world is mind-stuff'. I think so through the sequence of derivations from Whitehead:

1. Reality is organism, that system of distinct entities whose mutual conditioning creates a whole which is more than the sum of the parts.

2. The model for organism is combination mathematics.

3. Combination mathematics is equally the model for literacy and meanings whose 'whole is greater than the sum of the parts'.

Thus the essence of both Eddington-Jeans and of Whitehead is the reality of *literacy*, the art of linguistic combination arrangements. Overall I think the conclusion from the 1930 Cambridge Club is

'Reality is organised mind-stuff.'

# PART TWO
# THE SYMMETRICAL PARADOX

Sir Arthur Eddington set the seal on the views of the 1930 Cambridge Club with his statements:

> Religion first became possible for a reasonable man of science about the year 1927

and

> The idea of a universal Mind or Logos would be, I think, a fairly plausible inference from the present state of scientific theory: at least it is in harmony with it.

It was a philosophical breakthrough and opened the way to a reconciliation between science and religion. But it died a natural death as the 1939–45 war swallowed the world of science, forcing it to pursue military aims.

## The author's intent

It is my purpose in writing this book partly to revive memories about that most hopeful philosophy based on science, but also to take it further and show that in the 1980s the case is even more convincing.

## The ground to be covered

But the requisite material is so rich and widespread that it is first necessary to review certain major views of science in order to determine the nature of the overall argument involved. In particular it is necessary to consider:

1. the status of the major law of physics, the Second Law of Thermodynamics, which would tell us that creation is moving towards its own extinction;

2. the credibility of Darwin's Theory of Evolution which stresses the accidental nature of the evolution of species.

The conclusions reveal a rather remarkable SYMMETRICAL PARADOX.

# 7

## THE SECOND LAW OF THERMODYNAMICS — THE WORLD OF ACCIDENT STATISTICS

In the first place I would make it clear that I shall not attempt to disprove the Second Law of Thermodynamics. As Sir Arthur Eddington stated:*

> The law that entropy always increases — the Second Law of Thermodynamics — holds, I think, the supreme position among the laws of Nature. If someone points out to you that your pet theory of the universe is in disagreement with Maxwell's equations — then so much the worse for Maxwell's equations. . . . But if your theory is found to be against the Second Law of Thermodynamics I can give you no hope; there is nothing for it but to collapse in deepest humiliation.

### What is entropy?

Entropy is the measure of the extent to which a system has come into a certain state under random ('accidental') forces. The simplest analogue is that of a pack of playing cards which were originally arranged in their four suits and sequences. If we now shuffle the cards then we increase their entropy, their state of random disorder, until they come into a state where no further shuffling can increase the random disorder, in which case we have reached maximum entropy. Two matters should be noted:

1. Entropy can be reduced by *conscious sorting*. We can lay the cards out on the table and collect them back into their original suits and sequences. Clerk Maxwell was the first to point out this

---

*The Nature of the Physical World, p. 81.

possibility that 'a conscious sorting demon' can reduce entropy and reverse the Second Law of Thermodynamics.

2. The shuffling of the card pack essentially takes place in a state of 'absent-mindedness', i.e. the mental opposite of the conscious sorting of (1) above.

## The Second Law of Thermodynamics

The Second Law of Thermodynamics is a generalisation of such concepts about entropy and can be stated:

If a closed system is in a configuration that is not the equilibrium configuration, the most probable consequence is that the entropy of the system will increase with time.

Put another way, if one had a half-shuffled pack of cards, then time provides further shuffling to increase the entropy or randomness. The law refers to thermodynamics because it is mainly applied to the motion of atoms in gases under the influence of heat, but it is equally applicable to the state of order in any system including pure information systems (which is what a pack of cards is).

## The philosophical significance of the Second Law of Thermodynamics

The great philosophical significance of the Second Law of Thermodynamics is that it can be equated to *the direction of time* as from past to future. A system of entities left to itself but capable of inner accidental collisions (i.e. shuffling) will increase its entropy with the passage of clock-time.

The derivation from this is that if we consider the universe to be a similar system of separate entities such as atoms, and if these atoms can collide with each other as shuffling, then the entropy of the universal system is increasing with time and *the universe is moving in the direction of increasing disorder*, i.e. the universe is running down with the passage of time. Indeed, the chapter from Eddington's book from which I extracted the quotation at the beginning of this chapter is headed 'The Running Down of the Universe'.

## The philosophical paradox

Thus we here encounter the germs of a Paradox which will feature largely in this book:

1. If the universe is a system of blind chance with accidental collisions ('shufflings'), then entropy is increasing with time and the universe is running down to effective extinction.

2. To set against this we have the concept that 'the stuff of the world is mind-stuff' and thus there could be an aspect which is not blind chance and accidental and so the universe could be sorting rather than shuffling, or at least there could be some sort of equilibrium between the two.

The analogue of the Paradox is that of a clockwork clock which is certainly running down with time but somebody may have a key and be winding it up at the same time. So we see the immense philosophical significance of the views of the 1930 Cambridge Club. The Older Physics from the nineteenth century associated with such names as Boltzman was immersed in mechanical concepts of the universe, and considered that a good scientist could make a mechanical model of all his problems (such as the structure of atoms). In that case the philosophical scene was set for a pessimistic outlook for humanity located in a system in which the future pointed to a running-down condition of increasing disorder, so one day 'the clock will stop'. To set against this was the so-called Modern Physics created by Planck, Einstein and Heisenberg who told us that the universe was not mechanical, that you could not make mechanical models of it, and that although accidental chance and shuffling might have some part to play it was by no means the whole story. So in due course the 1930 Cambridge

Club erected another point of view that reality is organic mind-stuff, 'sorting stuff'.

## The newer ideas about entropy

In recent years there has been an interesting development which tends to separate entropy from its relationship to the Second Law of Thermodynamics and to associate it with general probability and improbability. This new interpretation can be defined:

The universe appears to consist of quantised parts such as atoms, but each such entity may have a number of distinct valid alternative states. The entropy of the system is equal to the number of such valid alternative states.

*Example.* Consider a system consisting of ten pennies each of which can be heads-up or tails-up. Toss them one at a time and lay them out in a row (imagine they are as A of Fig. 7.1). Now repeat the process and imagine they are now as B of Fig. 7.2. What this Illustrates is:

Each tossing reveals a *specific* arrangement.

The successive tossings are specific but different.

A little mathematics reveals that there can be $2^{10} = 1,024$ such alternative specific arrangements, and so the entropy of the situation is said to be 1,024.

On this analysis the entropy of a system is stated to be the number of its possible arrangements, provided each is valid and without restrictions. To such a system the normal mathematics of probability apply. This new statement of entropy gives exactly the same mathematics as the older variety based on the Second Law of Thermodynamics.

But why does entropy increase with the number of possible arrangements in a system of entities? Because entropy is the measure of the *improbability* of entities being in one pre-decided state; it is a measure of uncertainty, the uncertainty of *specificity*.

A

Fig. 7.1.  First row of ten heads-tails.

B

Fig. 7.2.  Second row of heads-tails.

Entropy is the probability that the next set of tossing ten pennies will *not* come down in a forecast pattern.

## The 'Law of Probability' as a misnomer

There has been great confusion among mathematicians as to whether there really is such a matter as the Law of Probability, i.e. a law which tells you how *likely* something is to happen. At the present time the experts take the view that it is only legitimate to consider that there may be a Law of Improbability, a law concerning the *unlikelihood* of something happening. Therefore the better definition of entropy is

Entropy is the measure of the unlikeliness of a specific pattern occurring from alternative possibilities.

Here we can go back to that card-shuffling exercise, and given a shuffled pack its entropy is the unlikelihood of then dealing out a pre-stated hand.

## Specificity

Later in the book we shall be much concerned with *specificity*, which can be described in two ways:

1. Specificity is the measure of the improbability of a pattern *which actually occurs* against a background of alternatives.

2. Specificity $= \dfrac{1}{\text{Entropy}}$

*The entropy with a pack of cards.* Let us imagine there is a pack of 52 cards well shuffled and lying face-downwards on a table. What are the chances of picking all the cards up in the correct suit sequence starting with (say) the Ace of Spades and working downwards and then through the other suits and finishing (say) with the Two of Clubs? Well the chance of picking up the first card correctly is 1 in 52, the second card 1 in 51, the third card 1 in 50, the fourth card 1 in 49 and so forth. So the chance of picking up the whole pack correctly is Factorial 52 (i.e. 52!) as:

$$\text{one chance in } \frac{1}{52 \times 51 \times 50 \times 49 \times 48 \times 47 \times 46 \times 45 \times 44}$$

$$\times \frac{1}{43 \times 42 \times 41 \times 40 \times 39 \times 38 \times 37 \times 36 \times 35}$$

$$\times \frac{1}{34 \times 33 \times 32 \times 31 \times 30 \times 29 \times 28 \times 27 \times 26}$$

$$\times \frac{1}{25 \times 24 \times 23 \times 22 \times 21 \times 20 \times 19 \times 18 \times 17}$$

$$\times \frac{1}{16 \times 15 \times 14 \times 13 \times 12 \times 11 \times 10 \times 9 \times 8}$$

$$\times \frac{1}{7 \times 6 \times 5 \times 4 \times 3 \times 2}$$

$$= \frac{1}{8.065811 \times 10^{67}} \text{ i.e. (about) 1 in } 10^{68}$$

This number is approaching that of all the atoms in the universe.

*The specificity with a pack of cards.* Now let us imagine that my reader is a magician and, faced with the same problem of picking up 52 face-downwards cards in the right suit and numerical order, does so first time. The *specificity* is the inverse of the above and is Factorial 52 ($52 \times 51 \times 50 \times 49 \dots$ etc.) and is thus odds-on of $10^{68}$. Thus if my reader can pick up all the cards correctly he, or she, is a $10^{68}$ magician.

Thus we note the mathematical symmetry between entropy and specificity:

$$\text{Specificity} = \frac{1}{\text{Entropy}}$$

## The symmetrical paradox between entropy and specificity

The polar or inverse symmetry as between entropy and specificity is of great philosophical interest since it shows that the Paradox as between the running-down of the universe and its winding-up depends upon the same general mathematics with an inverse or NOT relationship.

We must agree with Eddington that the Second Law of Thermodynamics is a major law of nature. But we find that it is only half the likely truth and that it has a complement in a sort of *Law of Specificity* which is its obverse using the same general mathematics. Gradually in this book we approach such a possible *Law of Specificity* and, perhaps to our surprise, we shall find it to be the central law of organism.

# 8

## DARWIN'S THEORY OF EVOLUTION

After the Second Law of Thermodynamics it is possible that science is next most committed to Darwin's Theory of Evolution as a major anchor-point. There is some correspondence since both relate to the domination of random and accidental influences. Darwin took the view that Natural Selection was such a powerful influence that a given species could evolve to being a different species. He also considered that 'in the beginning' there were only a handful of species, perhaps even just one, from which all others had evolved by Natural Selection.

### Brief description of Natural Selection

The Darwin hypothesis was:

1. Creatures are subjected to accidental variations (we should now say 'in the genes'), and when these are favourable for life survival, then such favoured creatures will do better in life and thus be selected by life's conditions to prosper.

2. But at the same time the effect of 'the use and disuse of parts' is a distinct separate evolutionary influence.

Simplistic Darwinism only takes (1) into account, in spite of Darwin's protests of the importance of (2), and he wrote:

Great is the power of steady misrepresentation.

For practical purposes concerning general opinion we can limit ourselves to (1) above, Simple Darwinism, but the problem is to account for those accidental variations (apart from sexual selection which appears legitimate and whereby the female mates with the male who is more dominant than the rest). People have thought

42

that accidental variations (we should now call them genetic mutations) could be caused by cosmic rays of various sorts. It is true that
mutations can be artificially created in the laboratory by heavy
doses of X-rays, but we also know that the natural strength of
such influences in the environment is far too weak to do the same.
It is now considered that any mutations must be of thermal origin
but, as Erwin Schrödinger has shown, such will be very rare
events.

As Schrödinger states:

> Granted that we have to account for the rare natural mutations
> by chance fluctuations of the heat motion, we must not be very
> much astonished that Nature has succeeded in making such a
> subtle choice of threshold values as is necessary to make
> mutation rare. For we have . . . arrived at the conclusion that
> frequent mutations are detrimental to evolution.*

Thus the effect of Natural Selection is not so much 'the survival
of the fittest' as 'the elimination of the unfittest', which is not
quite the same thing. It would seem that mutations are not part of
Nature's plan.

## What did Darwin actually say?

There is great misunderstanding as to what were Darwin's ideas in
*Origin of Species*† In the first place Darwin never attempted to deal
with the problem of the origin of life:

> There is grandeur in this view of life, with its several powers,
> having been originally breathed by the Creator into a few forms
> or into one and that. . . . from so simple a beginning endless

---

*What is Life?* (Cambridge University Press, 1967), p. 68.
†The quotations in this chapter are taken from the final chapter or 'Conclusion'
of his book.

forms most beautiful and most wonderful have been and are being evolved.

Thus Darwin held to a religious view concerning the origins of life, and he was only concerned with how different varieties came about. Although the title of the book refers to 'species', this was not a favourite word of Darwin's; he preferred to consider everything as *varieties*. Darwin's theory of evolution was concerned with the likelihood that chance variations in the make-up of a creature would have more or less survivability according to environmental conditions so that the environment acted like a filter which favoured certain changes, and this he called Natural Selection. It makes good sense. But he never stated that this was the only evolutionary factor:

> . . . my conclusions have lately been much misrepresented, and it has been stated that I attribute the modification of species exclusively to natural selection. I may be permitted to remark that in the first edition of this work, and subsequently, I placed in a most conspicuous position — namely, at the close of the Introduction — the following words: I am convinced that natural selection has been the main but not the exclusive means of modification. This has been of no avail. Great is the power of steady misrepresentation.

So what other factors did Darwin consider as relevant as well as natural selection?

> . . . natural selection . . . aided in an important manner by the inherited effects of the use and disuse of parts . . . and by the direct action of external conditions, and by variations which seem to us in our ignorance to arise spontaneously. It appears that I formerly underrated the frequency and value of these latter forms of variation, as leading to permanent modifications of structure independently of natural selection.

At the same time Darwin did not believe in the succession of acts of creation for species:

> . . . do they really believe that at innumerable periods in the earth's history certain elemental atoms have been commanded suddenly to flash into living tissues?

And

> A celebrated author and divine has written to me that he has gradually learnt to see that it is just as noble a conception of the Deity to believe that He created a few original forms capable of self-development into other and needful forms, as to believe that He required a fresh act of creation to supply the voids caused by the action of His laws.

From the above it would seem that Darwin considered that the Creator produced an original set of phylogenic prototypes such as:

Prototype bacteria
Prototype insects
Prototype plants
Prototype fish
Prototype animals
    etc. etc.

and then left evolutionary forces related to the environment to produce the variety ramifications. But Darwin had a personal preference, about which he was not dogmatic, for there being a *single* original prototype. Oddly enough, the information capacity of the DNA is of such magnitude as to be able to include 'all possibilities' so that even man could be potential in the DNA of the earliest forms of life such as protista. There is very little difference in the design of a uni-cell protista and a uni-cell sperm cell. Both have DNA as the organising element. Thus if there is a fundamental enigma as to the creation of life it centres on the DNA, not as a mechanism but as organic information of astronomical extent.

## Lamarckism

It is commonly held that Darwinism is opposed to an alternative, Lamarckism. The only reference in Darwin's book to Lamarck is very complimentary although Darwin does not deal with Lamarckism in depth. The essential view of Lamarck was that species were not permanently fixed (Darwin took the same view ) and that the main evolutionary influence comes from the environment (we have just noted Darwin taking exactly the same view as an addition to Natural Selection). Lamarck stated that the environmental changes create a new *need* and that the creature's *mind* moves behaviour towards satisfying that need and the revised behaviour is then inherited.

There was not an enormous difference between the views of Darwin and Lamarck after Darwin had admitted the importance of the 'use and disuse' of faculties and accepted their inheritance.

## Chapter summary relevant to the Paradox

Darwin assumes that geological times are long enough to provide opportunity for major modifications of a species so that it could be transformed into different species. This we shall refute.

# 9
## 'NOTHING BUT OR SOMETHING MORE'?

Let us now move into the No Man's Land separating the two aspects of that Symmetrical Paradox: the paradox as between entropy and specificity. Jacquetta Hawkes, an archaeologist and author of distinction, wrote on the subject in her Danz Lecture to the University of Washington at Seattle. The title of her lecture was 'Nothing But or Something More'.

> In 1954, in a book called *Man on Earth*, I was bold enough to say that I could not believe that the evolution of life on Earth was exclusively the result of natural selection working upon random variation. Soon afterwards I was hit by a broadside from Julian [Sir Julian Huxley, the distinguished biologist and descendant of Thomas Huxley, a Darwin protagonist]. He wrote, to summarise, that I showed unforgivable ignorance, arrogance and impertinence in my puerile questioning of Neo-Darwinism. Its tenets, he explained, had now been proved mathematically and stood far more securely than the pillars of the British Museum.
>
> Later on, however, when the Huxleys were staying with us, we had an amicable discussion, and after I had managed to produce one or two of the more scientifically based difficulties in the doctrine, his whole attitude changed. He paced up and down my room in a state of emotion asking: 'But if it is not natural selection working on random variation, what can it be? What can it be?'

### Reductionism ('Nothing But')

Reductionism (or 'nothing but') is that attitude of mind which only feels it can keep in touch with reality if it narrows its outlook to specific facts, and preferably facts which can be interpreted in numbers or what Eddington called 'pointer readings'. Jacquetta Hawkes further states:

> The analytical approach has had such astounding successes in the physical sciences that it has produced an equally outstanding hubris among the smaller-minded scientists. What was really a method, one way of turning our brains upon limited aspects of the universe that has produced them, has tended to become a view of life, a totalitarian ideology. It has been said that nothing that cannot be measured and proved experimentally has any validity. . . . Reductionist thought suggests that the whole is no more than the sum of its parts and so leads to an old-fashioned mechanistic view.

Let us be quite clear about the limitations of Reductionism. What it does is to abstract from Actuality (see Chapter 2) and consider only those factors which are both 'fixed' and to which numbers can be assigned. Thus the Reductionist believes in:

The velocity of light
Avogadro's number
Planck's constant
The mass of neutron
The charge on the electron.

I also believe in these things but they leave out almost the whole of Actuality including:

The nature of human beings
Biology and life
Art, philosophy and religion.

Even within the field of physics the Reductionist point of view cannot account for:

Relativity
Quantum physics

The Uncertainty principle
The Exclusion principle.

As Jacquetta Hawkes says, the 'nothing but' attitude of Reductionism leads only to a cul-de-sac of old-fashioned mechanistic views and their corresponding outdated (outdated since 1900) mechanical models of reality.

## Organicity ('or Something More')

No one will question the validity of Reductionism within its proper sphere of detail analysis, but it belongs to that outlook which 'cannot see the wood because of the trees' and while the trees are real so is the wood. The theme of organicity in which the whole may be more than the sum of the parts is not opposed to Reductionism but is its complement. In terms of the philosophical theme being developed in this book, Reductionism and Organicity are also two equivalent complementary aspects of that Symmetrical Paradox. Jacquetta Hawkes:

> Against the basic methods of Reductionism one has to set the universal reality of hierarchy. From molecular chemistry upwards existence is structured in this mode, and at each level the parts are, as it were, two-faced, each organising the smaller ones below and being organised by the larger ones above. At each level, too, there emerge new properties not present in those below. Each organism, hierarchically controlled in this way, is able to maintain itself in space for an appreciable time. . . . In other words I stand before you as a specimen of that highly self-conscious type of organism *Homo Sapiens*. What makes this hierarchical system of control over billions of molecules more remarkable . . . is the fact that these molecules are in a state of continual flux, while the cells of my brain are at every moment dying in pale hosts.

## *The views of A. N. Whitehead and a word from Louis de Broglie*

It was A. N. Whitehead (Chapter 5) who went some way towards reconciling Reductionism with Organicism with his ideas about organic mechanism. Typical was the thought that a particle such as an electron 'blindly runs' — but according to the 'railway lines' of the organic system in which it finds itself. So also when an electron is part of an atom its behaviour comes under the systematic rules of quantum mechanics.

The great French physicist Louis de Broglie (who forecast Wave Mechanics) also wrote significantly:*

. . . Quantum Physics. This has shown that there is a kind of complementarity between the concept of the individual unit and the concept of a system. In Quantum Physics, therefore, the system is a kind of *organism*, within whose unity the elementary constituent units are almost reabsorbed. When forming part of a system, then, a physical unit loses a large measure of its individuality, the latter tending to merge in the greater individuality of the system. . . . To make a real individual of a physical unit belonging to a system, then, it is necessary to take this unit from out of the system — to break the links which bind it to the total *organism*. If this is understood, we can also understand the way in which the concepts of the individual unit and of the system are complementary: the particle cannot be observed so long as it forms part of the system, and the system is impaired once the particle has been identified. The concept of the physical unit thus becomes completely clear and properly defined only if it is regarded as a unit completely independent of the rest of the world: but, since such an independence obviously cannot be realised, the concept of a physical unit taken in absolute strictness is found in its turn to be an idealisation; in other words, to be a case which is never rigorously adapted to reality.

---

*In *Matter and Light* (Eng. transl. London: Geo. Allen and Unwin, 1939).

This is a very interesting statement because it suggests the establishment of organism right within the central body of physics. This could be interpreted as the innate influence of organism (the complementarity between parts and wholes) as the basic structure in physics.

## Chapter summary relevant to the Symmetrical Paradox

The Symmetrical Paradox (see Chapter 7) tells us that probability (as measured by entropy) and improbability (as measured by specificity) are numerically inversely related. Such a pair of extremes are based on identical but inverted mathematics. This logical relationship appears to be very similar to, perhaps identical with, the structure of organism with the complementary relationship as between parts and whole. An organism which is specific as to its whole can thus be constructed from ultimate parts of only statistical stability. So, as Jacquetta Hawkes points out, the cells of the human body can die in millions without undermining the integrity of the whole. It is suggestive that the Symmetrical Paradox is an organic paradox.

# 10

## THE PROBLEM OF EVOLUTION — MONKEYS WITH TYPEWRITERS

We return to those words of Sir Julian Huxley:

> If it is not natural selection working on random variation, what can it be?

### The statistical approach

There is a recurrent line of thought among both lay people and non-mathematical scientists that, provided one had the right basic atoms and molecules to hand, then life could evolve by accidental aggregation and random mutations combined with the Darwinian principle of the survival of the fittest. To illustrate the point it has been suggested that if millions of monkeys were to type in random fashion, they would one day type a Shakespeare sonnet. *It is not true* and I took time off to do the random mathematics. Not being an expert on Shakespeare's sonnets I chose one verse of a poem I know well, Wordsworth's *Daffodils*.

> *I wandered lonely as a cloud*
> *That floats on high o'er vales and hills*
> *When all at once I saw a crowd,*
> *A host of golden daffodils,*
> *Beside the lake, beneath the trees,*
> *Fluttering and dancing in the breeze.*

The verse contains 159 letters and the following are the individual frequencies of occurrence:

| | | | | | | | |
|---|---|---|---|---|---|---|---|
| *a* | 17 | *c* | 4 | *e* | 22 | *g* | 4 |
| *b* | 3 | *d* | 11 | *f* | 5 | *h* | 10 |

| $i$ | 9 | $o$ | 11 | $t$ | 12 | $k, v, z, y$ 1 each |
|---|---|---|---|---|---|---|
| $l$ | 13 | $r$ | 6 | $u$ | 2 | |
| $n$ | 13 | $s$ | 9 | $w$ | 4 | |

The 159 letters can be permuted in the following number of ways, only one of which would be the verse *Daffodils*:

$$\frac{159!}{n_1! \times n_2! \times n_3! \text{ etc.}}$$

where $!$ implies factorial expansion such as 159 $\times$ 158 $\times$ 157 etc., and $n_1$ etc. are the frequencies of the different letters and are also to be factorially expanded. So if $n_1$ is the letter 'a' of which there are 17, then it has to be expanded as to 17 $\times$ 16 $\times$ 15 etc.

If one works out this equation it comes to about $3.6 \times 10^{175}$ so that those monkeys would need to type $3.6 \times 10^{175}$ verses if one of those verses were to have a high probability of being *Daffodils*.

Now to those monkeys:

1. Assume each monkey can type the 159 letters in a half-minute, quite a fast rate of typing. So a monkey working 24 hours a day can type about 1,000,000 'verses' in a year.

2. Let us have a troop of 1,000,000 monkeys all typing away together.

3. Let them type away for 1,000,000 years.

At the end of that time they will have typed $10^{18}$ random verses compared with the $3.6 \times 10^{175}$ required for *Daffodils*. The shortfall is a factor of $3.6 \times 10^{157}$ and those monkeys would need to have typed for $3.6 \times 10^{163}$ years. This is rather a large number:

3.6 $\times$ 10000000000000000000000000000000000000000000000000000000-
0000000000000000000000000000000000000000000000000000000000-
0000000000000000000000000000000000000000000000000000 years

Not all the monkeys there have ever been, typing for the span of organic life on Earth, could make the slightest impression on the problem.

In the previous chapter we had a description by Jacquetta Hawkes of how she brought certain data to the attention of Sir Julian Huxley to support her case that life could not have arisen by a combination of random chance variations combined with Natural Selection. I do not know just what evidence she produced, but I imagine it might be something like the foregoing calculation with typing monkeys which demonstrates the astronomical escalating odds against the random shuffling of letters to create meaningful statements.

## *T. H. Huxley and typing monkeys*

It was only after I had written this chapter to the present point that I came across a reference to the origin of ideas about 'typing monkeys'. This was in *The Mysterious Universe* by Sir James Jeans (p. 4) where he states:

> It was, I think, Huxley who said that six monkeys, set to strum unintelligently on typewriters for millions of millions of years, would be bound in time to write all the books in the British Museum.
>
> If we examined the last page which a particular monkey had typed, and found that it had chanced, in its blind strumming, to type a Shakespeare sonnet, we should rightly regard the occurrence as a remarkable accident, but if we looked through all the millions of pages the monkeys had turned off in untold millions of years, we might be sure of finding a Shakespeare sonnet somewhere amongst them, the product of the blind play of chance.

ALAS, NO! Both Huxley ('all the books in the British Museum') and Jeans ('a Shakespeare sonnet') had it quite wrong. But let us proceed to the proof (forecast in the analysis of the poem *Daffodils* above) before we draw ultimate conclusions as to what chance can and cannot do.

The reader will recall that back in 1860, after Darwin had published *The Origin of Species*, his chief supporter was T. H. Huxley who was a qualified surgeon and a noted educationist. Indeed Huxley somewhat took Darwin's place since Darwin was back-tracking (see Chapter 8) as to the unique validity of his theory of evolution and so Huxley took the prominent place as the advocate of this theory. At the memorable encounter at Oxford between Huxley and Bishop Wilberforce (a fundamentalist) it is the common opinion that Huxley 'floored' the Bishop. Since that time Darwin's theory of evolution has been the accepted wisdom of the world of science.

Now both Darwin and Huxley considered that *improbable chance*, however improbable but *possible*, supported the theory of evolution, and we have given Huxley's analogue of typing monkeys above. But how true was that analogue? This we explore next.

*Thesis.* That six monkeys typing randomly for 'millions of millions' of years would type all the books in the British Museum.

*Relevant facts*

1. In 1860 there were 700,000 books in the British Museum.

2. If each book is of 350 pages with 40 lines to the page there would be $9.8 \times 10^9$ lines of type in all the books.

3. It can be shown that with 50 letters per line the random permutations of a line of print will be about $50!/17! = 8.5 \times 10^{49}$ (the $17!$ is an allowance for the recurrence of identical letters, derived by sampling). In other words, a typical line of bookprint can be arranged in $8.5 \times 10^{49}$ ways.

4. Allowing for the multitude of lines of print in the books (at $9.8 \times 10^9$ as above) then the improbability of randomly typing one given line in all the books is:

$$\frac{8.5 \times 10^{49}}{9.8 \times 10^{9}} = 8.6 \times 10^{39}$$

5. A monkey typing at human rates could type 5 lines of print in one minute so six monkeys could type 30 lines of print in one minute, $1.6 \times 10^7$ per year.

6. As to 'millions of millions' of years let us take this as the estimated life of the universe at $10^{10}$ years. So the monkeys could type $1.6 \times 10^7 = 10^{10} = 1.6 \times 10^{17}$ lines.

So we see that Huxley's monkeys could type $1.6 \times 10^{17}$ lines against an improbability of $8.6 \times 10^{39}$, so the shortfall would be the factor of $5.4 \times 10^{22}$. Now Huxley only nominates six monkeys, so let us give him a generous allowance of monkeys at something about equal to the human population in 1860, say $10^9$. The shortfall would still be a factor of more than $10^{14}$. So:

ALLOWING HUXLEY ALL THE MONKEYS THERE HAVE EVER BEEN, TYPING FOR ALL THE TIME THERE HAS EVER BEEN, THERE WOULD BE A SHORTFALL RATIO OF MORE THAN ONE HUNDRED MILLION MILLIONS, AND THAT ONLY RELATES TO THE CHANCE OF TYPING ONE LINE OF ONE BOOK IN THE BRITISH MUSEUM.

## *Just how much could Huxley's monkeys have typed?*

The final truth is to know just how much Huxley's monkeys could have typed. It is calculable, and it is obviously less than one line of print:

1. The monkeys can type $1.6 \times 10^{17}$ relevant lines (relevant to any line of print in any of the books).

2. The shortfall relevant to a line of 50 letters is the factor $5.4 \times 10^{22}$.

3. By shortening a line of print to fewer than 50 letters (*a*) and (*b*) balance at a line of 36 letters.

So:

Huxleys six typing monkeys typing for the duration of the universe would type 36 letters of sense in one of the books in the British Museum.

Hardly 'all the books in the British Museum'; and (vide Jeans) only part of one line of a Shakespeare sonnet. Huxley had it wrong, and inasmuch as Darwin relied on the same sort of improbability argument, then Darwin had it wrong too.

But how wrong! By factors of millions of millions. And if one brought all the books in the British Museum into the calculations, then it would be wrong by millions of millions of millions of millions. Of course they did not have computers in 1860. But they did have logarithms which could serve the same purpose.

## *What is it?*

So we are back to Jacquetta Hawkes and Sir Julian Huxley (Chapter 9) wondering:

But if it is not natural selection working on random variation, what can it be? What can it be?

# 11

## REFLECTIONS ON THE SYMMETRICAL PARADOX

This book began with the views of the 1930 Cambridge Club who had come to the conclusion that the nature of ultimate reality was organic mind-stuff. In this Part 2 we have considered two of the main ideas of conventional science (conventional in the sense that they were the main ideas of the nineteenth century although still with us) as to the Second Law of Thermodynamics and also Darwin's Theory of Evolution.

The Second Law of Thermodynamics tells us that disorder increases with time in 'blind' mechanical systems subjected to internal collisions such as between gas molecules, which we detect as temperature. This disorder is known as entropy and is a valid concept provided we realise that it only applies to certain circumstances of 'blind collisions'.

On Darwin's Theory of Evolution we found by calculation (those monkeys with typewriters) that there had not been sufficient time since universal creation for evolution to have occurred through a random chance element. Accordingly we have to invent a new name to describe the nature of evolution, and the most appropriate name is *specificity* (the choice of name is not surprising related to *species*).

If we consider entropy it is known that this refers to the behaviour of 'crowds' whereas specificity refers to individuals, and the question is whether there is a relative process which links the two together. In other words, are entropy and specificity two aspects of the same thing? for they are certainly the two ends of the same stick of 'relative order'.

## It is a paradox

Let us consider the essentials of the two aspects:

1. There is a process of running-down from order to chaos, applying to crowd situations and measured by entropy.

2. There is a process of winding-up from chaos to order, applying to organism and measured by specificity.

As to the second point Erwin Schrödinger has written about organism:

> It feeds upon negative entropy, attracting as it were, a stream of negative entropy upon itself, to compensate the entropy increase it produces by living . . .*

Thus I conclude that it really is a paradox since one does not expect to find processes of running-down and winding-up going on together. The philosophical problem of the older physics is that it has tended only to observe the running-down process and so it had to invent the Big Bang to establish a potential of order from which running-down could descend. The fact appears to be that there really are two distinct processes which are not two stages of the same process of relative order. A specific process is not one with 'less entropy', but it is one without any entropy at all! We note this in specific organisms such as animals: they are either alive or dead but one does not find them in the half-and-half state.

There is something very strange about specificity, a sort of magic. Earlier we referred to the chances of picking up all the cards of a pack in the correct order (or dealing four hands each of one suit only) and noted that the odds against were astronomical and that if anyone could do this at will he would be a magician!

## The search for specificity

Let us now abandon this pre-occupation with the Symmetrical Paradox and replace it with a single mystery — the subject of specificity. To study specificity one can hardly do better than turn to molecular biology.

*What is Life?*, p. 78.

# PART THREE

# SPECIFICITY AND MOLECULAR BIOLOGY

It is reasonable to suppose that organic creatures are specific since each is a unique individual. But the recent developments in molecular biology, including the cracking of the DNA code, enable us to put fairly accurate figures on to the degrees of specificity involved. These turn out to be astonishing. The DNA is an information complex the equivalent of 20,000 long books while the improbability of the haemoglobin molecule is represented by an odds-against number exceeding $10^{650}$.

I would make it clear that I am not a molecular biologist, and that my deductions should be considered in terms of principles rather than nut-and-bolt exactness. For example, in the section of Chapter 13 entitled *The elegance of DNA literacy* I suggest letters representing pseudo-codons, but they could well be different letters. The existence of pseudo-codons which do not code for specific amino-acids is now well established. (An article by Stebbins and Ayala called 'The Evolution of Darwinism' in the July 1985 issue of *The Scientific American* shows that the DNA of the amoeba is longer than that of man, but most of it consists of millions of repeating pseudo-codons which do not code for anything but which are not 'nothing'. It is as if the spare DNA spaces were filled with asterisks to ensure that they cannot be used for chance meaningful codons.)

# 12

## SPECIFICITY IN PHYSICS AND CHEMISTRY

In this Part 3 we are concerned with specificity in Nature. Let us contemplate whether specificity is built into Nature 'from the ground up'. Thus we contemplate a hierarchy of specificity starting with the finest and smallest 'materiality' — that of radiation such as light.

### The specificity of radiation

It was Max Planck at the beginning of this century who discovered that radiation was specific in that it consisted of packets or 'bullets' which obeyed the law:

$E = hF$ where $E$ was the energy of each 'bullet'.

$F$ was the wave frequency.

$h$ was a constant, now called Planck's Constant.

The specificity resides in the exactness of Planck's Constant, which has the value $6.62 \times 10^{-27}$ erg-seconds. Now this is a very extraordinary fact since one would expect that the more energy was available at the source of radiation such as light, then the more energy would be radiated on a linear increasing scale. But not so: the energy is chopped up into standardised 'bullets' whose size is determined basically by Planck's Constant.

### The specificity of all the main laws of physics related to the specificity of their associated cosmic constants

The above illustration of specificity related to radiation is only one of a whole number of basic laws of physics which are specific because of associated specific constants. Altogether there are six fundamental atomic constants:[*]

> Avogadro's number
> The electronic charge
> The electronic rest mass
> Planck's Constant
> The velocity of light
> The mass of the neutron.

From these six constants, a total of about sixty other derived constants can be calculated covering almost the whole scope of physics.

Let us not go into the detail of all this but simply conclude:

THE WHOLE OF PHYSICS, AT THE GROUND FLOOR LEVEL, IS DOMINATED BY SPECIFICITY QUANTIFIED BY THE ATOMIC CONSTANTS.

This is a conclusion of profound philosophical interest. The charge on the electron is exactly the same measured in a laboratory on our planet Earth or measured in the Spiral Nebulae of Andromeda, even though we have never been there! This is not some statistical averaging but applies to the entities of physics taken 'one particle at a time'.

## The specificity of atoms

Radiation such as light is the 'fine structure' of physics, and so we next move up the hierarchical ladder to the atom where a new factor of specificity comes into play, the specificity of whole (integral) numbers. For this we also have to acknowledge the genius of Max Planck — combined with a nod to the Russian scientist Mendeleef of the nineteenth century.

[*]Condor and Otishaw, *Handbook of Physics* (New York: McGraw-Hill), Chapter 10: 'Fundamental Constants of Atomic Physics'.

I think it true to say that we human beings know of nothing more specific than whole numbers. I have three children and while my judgment of them as to their nature and characters may vary almost with the time of day, I cannot doubt that I have specifically *three* children and not two or four.

The atom is specified by the number of electrons in its outer orbital shells. The main atoms which create organic materials are:

| Atom | Number of electrons in outer shells |
|------|------|
| Hydrogen | 1 |
| Carbon | 4 |
| Nitrogen | 5 |
| Oxygen | 6 |

## The specificity of molecules

Quantum mechanics tells us that atoms will link up with other atoms of a different kind when the total number of electrons shared between the atoms is eight — the Rule of Eight. This is proved by the fact that those atoms which already have eight electrons in their outer orbits, such as neon and argon, are chemically inert. They have no 'electronic room' to share with other atoms. The result of this is that atoms tend to make molecular compounds with other atoms according to the Rule of Eight, and the principal result significant for organic molecules is as follows:

*Water.* Two atoms of hydrogen will link up with one atom of oxygen to form water $(1 + 1 + 6 = 8)$.
*Carbon Dioxide.* One atom of carbon will link up with two atoms of oxygen to form carbon dioxide $(4 + 6 + 6 = 16,$ but $16 - 8 = 8)$. Note that in this case the final chemistry moves to the second outer 8-shell for its resolution.
*Ammonia.* One atom of nitrogen will link up with three atoms of hydrogen to form ammonia $(5 + 1 + 1 + 1 = 8)$.

The above three molecules are the prime raw material for

organic life. Water and carbon dioxide are the basis for carbo-hydrates, and the addition of ammonia creates the possibility for amino acids (for proteins).

## *Hierarchical numerical specificity of physics and chemistry*

From the above brief analysis we note that there is a hierarchical specificity as from physics to chemistry (I have only dealt with the ladder to organic chemistry) according to the following stages:

1. The cosmic constants related to radiation and atomic physics.

2. The above establishes the numerical exactitude of prime entities such as electrons and neutrons, the raw material for atoms.

3. Atoms are defined by the number of electrons in their outer orbits.

4. The propensity of atoms to join up with each other is deter-mined by the Rule of Eight (i.e. covalent bonds) and leads to the following organic chemicals as occurring naturally from purely physical numerical laws:

> Water
> Carbon Dioxide
> Ammonia.

These three compounds are the prime raw materials for life.

## *The philosophical significance of numbers*

In considering the above four-stage system whereby the cosmic constants eventually emerge as the prime organic materials for life, we are essentially in the realm of *numbers* — the number of the cosmic constants, the number of electrons in the outer orbits, the Rule of Eight. Physics knows quite a bit about all this (particularly from Quantum Theory), but it does not have fundamental explanations. But from a broader philosophical point of view one can conclude:

THE UTTER BASIS OF THE UNIVERSE IN GENERAL, AND ORGANIC LIFE IN PARTICULAR, IS NUMBERS.

It is a mystery. But whatever the explanations, it is clear that one is in the realms of utter *specificity*, for there is nothing more specific than fixed numbers.

## Organic chemistry

From the previous section we have seen how the prime 'foods' for organic life (water, carbon dioxide and ammonia) are a natural outcome of the numerical properties of physics, thus establishing a basic ladder between physics and life. And we recall that these numerical findings are still a 'mystery'.

Let us consider the destination of just two of these organic 'foods', water and carbon dioxide. The first thing we note is that these two valuable materials have descended to the bottom of the energy ladder with their collective eight electrons and thus are reluctant to enter into new chemical processes. They are as fixed and settled in their ways as the noble gases such as neon or argon. They could stay just as they are through all eternity unless they were supplied with external inputs of energy to break up their passive inertia.

## Bioenergetics

But where is the external energy to come from to rouse water and carbon dioxide from their slumbers? It comes from the Sun in the process of photosynthesis whereby green vegetation converts sunlight into electricity, which in turn makes chemical energy available for that 'chemical factory' we know as a living creature. What happens is:

1. Photons of sunlight impinge upon the chlorophyll in vegetation, and the photon energy raises an electron to a state of high energy; this is the prime energy we are looking for to break up water and carbon dioxide.

2. But not directly, and the high electron energy is first put into a 'storage battery' rather like the way the electricity from our car generator is put into electro-chemical storage in our car battery. This natural battery is made from a chemical, ATP (adenosine triphosphate), and it is the phosphorus in ATP which is the chemical basis of energy storage (not unlike the phosphorus with which we tip the ends of matches so we can strike a light).

3. After the electron has given up its energy to the ATP battery, it falls back to its low-energy state to be ready for another energy cycle created by more sunlight photons falling upon the chlorophyll of green vegetation.

So overall the process is quite simple. Sunlight converts to electricity which is put into battery storage in the plant. But just one other point. The electrical energy stored in ATP is very high-class high potential energy, and I have likened it to the phosphorus on the tip of a match. But what is the 'wood' of the match which operates at a much lower energy level — that of mere heat as distinct from electricity? That lower energy level ('wood') is sugar, typically in the form of glucose. Thus the motor-car analogy is that phosphorus is the sort of energy suitable for the spark ignition system, while sugars are more like the petrol in the car tank. We shall come across this fact later in considering the structure of DNA.

## The need for structural organisation

So we know where the energy comes from to re-arrange that prime water and carbon dioxide: it comes from sunlight and it is transferred as electricity into ATP phosphorus electrical batteries. That is from an energy point of view. But obviously the whole process requires a mass of *organisation* in green leaf structure, in places where the ATP can get to work on water and carbon dioxide to make sugars and much more besides. Put shortly, what is needed is a *chemical factory*. We have accounted for some of the main *inputs* required by that chemical factory, inputs such as water, carbon dioxide and sunlight, but we have no clue as to the organisation of the chemical factory itself, the green-leafed plant. That organisation is provided from within the plant itself, from a 'managing director' called DNA — whom we had better now interview!

Whereas the energetical aspect of life is concerned with organic chemicals such as carbohydrates and sugars — materials which can 'burn' to create energy —, the structural chemicals of the body involve the incorporation of nitrogen (from that ammonia), and are most typically the proteins. It is the proteins which are the building bricks of *structure*, whether the structure be organs, muscles or blood.

# 13

## AMINO ACIDS, PROTEINS AND DNA

### The chemical factory

So far all we have established is that sunlight provides the ongoing energy for life through photosynthesis of electricity in green vegetation, and that water and carbon dioxide are two main materials as inputs to plants.

But now let us consider the whole of organic life, whether microbes, plants or animals (including man), and realising that each needs to have a chemical factory, which in the case of animals has to be a *mobile* chemical factory. If we consider a man-made chemical factory, say for the production of nylon fibres, it needs to have the following main functions:

1. It needs a building within which the chemical plant can be organised.

2. It needs a string of unit chemical process machines to subject the input materials to the several stages of manufacture.

3. It needs a conveying system to move materials from one stage of manufacture to the next.

If we now apply the above to man, we note that the 'building' is his body taken as a whole. The 'unit chemical process machines' are the organs of the body. The 'conveying system' is the bloodstream. In man all this structure is made from a set of standard building bricks which are (mainly) the *proteins* (that Greek word means 'chief'). So the proteins are the chief building bricks of the human body, and the problem is how they are created, for each is highly specialised. Although we eat protein aggregates such as beefsteaks, the variety of proteins which those contain cannot be used directly by the human being since they would not be of the correct species nature for the human body. So these ingested proteins are first broken down into their sub-building bricks known as amino-acids. Given amino-acids, then, the human body can rebuild from them proteins of the correct species form in amounts needed for human purposes.

## The amino-acid sub-building bricks

The amount of building-up required is illustrated in the following table:

| Material | Molecular weight |
|----------|------------------|
| Water | 18 |
| Amino-acid | *typically* 125 |
| Protein | *typically* 50,000 |

So we see that an amino-acid is about seven times as complex as water, and protein is about 400 times as complex as amino-acid. This implies that to make a typical protein would require about 400 amino-acid molecules all strung together and in a specific order. The alternative number of permutations is immense, 'countless trillions'. Therefore any manufacture of a specific protein requires very exact control, and this establishes a marked characteristic of molecular biology which is its *specificity*.

Let us now assume that we have had a meal of beefsteak and that this has been broken down by our digestion into a variety of amino-acids which pass into the bloodstream. An amino-acid has the chemical formula shown in Fig. 13.1 (overleaf). At the left end is the amino group which carries a positive electrical charge, while at the right end there is an acid group which carries a negative electric charge. Thus if the head of one amino-acid approaches the tail of another amino-acid, they will join up in strings due to the attraction between opposite electric charges as shown in Fig. 13.2. So we see that amino-acids have a natural propensity to join up in strings.

But also shown in Fig. 13.1 at the top is R (standing for

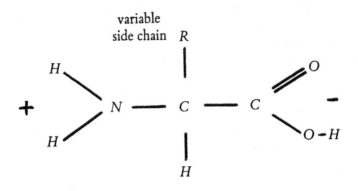

Fig. 13.1. Basic amino-acid.

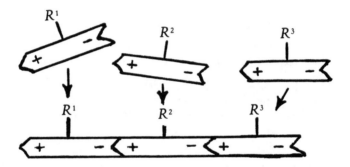

Fig. 13.2. Amino-acids join up head to tail due to electrical charges.

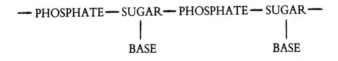

Fig. 13.3. DNA backbone of phosphate-sugars with side links of bases which determine the DNA code.

radical); this is the variable from one amino-acid to another and may vary from just a single hydrogen atom (as in glycine) to quite a complex ring structure (tryptophan). These variable side chains can be looked upon as *keys* which can fit only corresponding *locks*. There are only twenty different amino-acids which are relevant for the construction of proteins.

## The DNA

It has been known for hundreds of years that the body grows from cells, starting with one cell which then splits into two, which then each split into two again until ultimately, after about thirty such divisions, the body will have 1,000,000,000 cells. Thus, in the absence of some 'life spirit', bodily growth and structure are determined by the cells, going back to one cell. It is at the centre of that original cell and all later cells that we find there is a nucleus, and within the nucleus are chromosomes and genes which are the real governing authority.

In this century it has been found that the gene is DNA (deoxyribonucleic acid) and is a single molecule whose typical molecular weight is several millions (compared with hydrogen of atomic weight = 1). DNA is not manufactured in the body since it is inherited from our parents via the sperm-ovum chemistry, but it can be replicated into the cell division system so that all the cells of the body have DNA in the cell nucleus.

The DNA is a tube-like molecule consisting of two spirals with cross linkages, and is a progression of 'nucleotides', each of which consists of:

One of four alternative 'bases' which describe its characteristic 'letter'.

This is connected to a sugar.

The sugar is connected to a phosphate group.

The arrangement is shown in Fig. 13.3, and the repeating

theme is sugar-phosphate/sugar-phosphate/sugar-phosphate etc. etc. But to each of the sugars is attached one of the four alternative bases which create 'letters'. Note that, as described in the previous chapter on bioenergetics, it is the sugar-phosphate pair which provides the stick and tip of the 'energy match', so that the DNA has its own source of energy to form other molecules. The sugar-phosphate chain can be considered as the skeleton or backbone of the DNA molecule, while it is the side tacked-on bases of four varieties which are relevant to the DNA manufacturing capacity and 'information'.

## *The alphabet and syllables of the DNA*

The four alternative bases which can be tacked on to the DNA molecule are:

| *Notional letter* | *Chemical* |
|---|---|
| C | Cytosine |
| A | Adenine |
| G | Guanine |
| T | Thymine |

I will not describe the structure of the bases, but they are quite simple molecules whose typical molecular weight is 125, about the same as the amino-acids.

If one has a system of letters, then English grammar suggests that the first way by which they could become more meaningful would be in syllables.

What the molecular biologists have shown is that although the vocabulary has four letters (*C, A, G, T*), these group in threes to form 'codons' or what I have called syllables. Thus typical syllables are:

*CAG   TGA   ACT   GAT*   etc. etc.

It can be shown that altogether there are 64 permutations of three letters at a time from a group of four. But this number is more than is needed since the purpose of a syllable is to code for one out of only twenty amino-acids. The aim is to create proteins which are themselves strings of amino-acids. We noted earlier that a protein typically contains 400 amino-acids in a specific chain sequence, so that the DNA corresponding code must be something like:

| 1 | 2 | 3 | 4 | 5 | to 400 |
|---|---|---|---|---|--------|
| CGT | GAT | CAG | TAC | GTA | etc. etc. |

## DNA, RNA and protein

The DNA does not create amino-acids directly in this fashion for it uses an intermediate RNA (ribonucleicacid), which is very similar to DNA but a sort of negative. Where DNA has mountains, RNA has valleys and *vice versa*. So the DNA first codes for and creates RNA, which then floats off and in turn assembles the amino-acids in the correct order to link up into specific proteins. Hence the famous statement:

DNA makes RNA makes proteins.

This whole system is:

1. The human body has two inputs:
   (a) (mainly) animal protein;
   (b) inherited DNA.

2. The stomach digests the animal protein to break down into 20 amino-acids.

3. DNA programmes its daughter RNA so that it can arrange the selection from 20 amino-acids in a definite sequence to be human proteins.

## Why DNA (and RNA) are codes

We have seen earlier that amino-acids are polarised by positive electric charges at one end and negative electric charges at the other end, so that they can easily be linked in chains by attraction of electric charges of opposite nature (Fig. 13.2). There is no reason why a particular set of amino-acids should form one chain rather than another set, but the DNA-RNA system does arrange the amino-acids in specific sequences to form a specific protein. Thus the system is like having a set of alphabetical letters which can be strung together to make words and sentences but only in a sequence which makes meaning and sense. The sequences of the DNA as prime programmer are meaningfully arranged to make some material required by the body. It is true that the DNA is fixed in its sequence, but that is how it is inherited; the same messages are carried forwards from generation to generation. If a letter is changed through the means of bio-engineering, then the meaning changes, and that changed meaning is now inheritable. But overall the DNA is a *code*.

## The elegance of the DNA literacy

The DNA may typically have a string of letters such as the top line below, but the problem is: if these group into syllables (codons) of three letters, how are the syllables *separated*?

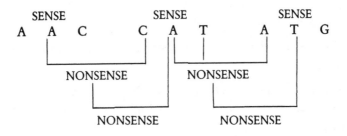

This is where those 44 surplus syllables come in as shown above, in that although they are represented in strings of three letters they are 'nonsense' and they do not code for an amino-acid. But they are effective as commas or full-stops. It can be shown that only 40 pseudo-codons can be employed in this fashion, one at each side of a syllable codon, and that leaves 44 − 40 = 4 unaccounted for. These are the pseudo-codons CCC, AAA, GGG and TTT which, having no inner specificity (i.e. different letters), are also 'nonsense'. Thus all 64 possible codons are accounted for as to

20 codons for 20 amino-acids

40 codons for comma or full-stop effects

4 unusable codons
——
64

The confirming proof is that Nature only uses 20 amino-acids. This was known before the above scheme was understood, and it then appeared as though Nature had 44 surplus codons. But these fitted exactly into the need for comma and full-stop functions as 'nonsense separation' together with the four meaningless triads of CCC etc. The DNA literacy is one of remarkable elegance.

## The DNA in every cell contains the total body information

The DNA is in the nucleus of every cell of the body, and since it is replicated from the original ancestral body cell, it contains all the information about the performance of *all* different kinds of cell and no doubt information relevant to higher degrees of integration of behaviour above the cell level. But the understanding in molecular biology is that the DNA of a given cell, in order to do its own work correctly, consists of the DNA which is 'switched on' to do that work, most of its information being switched off so as not to interfere with its proper duties.

Let us consider the general fact that every DNA molecule has the immense information corresponding to all workings of the body. This was demonstrated dramatically by J. B. Gurdon:[*]

*C. U. M. Smith, *Molecular Biology* (London: Faber and Faber, 1968), p. 384.

The essence of Gurdon's technique is to remove nuclei from unfertilised frog's eggs and to replace them by nuclei taken from the specialised cells of the frog tadpole. . . . Gurdon was able to show that it was possible to cause the eggs possessing transplanted nuclei to develop parthenogenetically into fertile male and female frogs. The specialised cells from which Gurdon obtained nuclei were fully developed intestinal cells. These cells were in every respect differentiated to carry out an absorbtive function. Yet Gurdon's experiment showed that their DNA complement was fully capable of programming the development of an entire organism.*

Thus the transferred nucleus which had most of its information switched off in its role of intestinal cell was switched on again when placed within the frog's egg, and one can only suppose that the switching-on was due to the influence of the new environment. This suggests that there may be some 'field' relationship as between a DNA nuclear molecule and its environment.

## The code system has never changed

The evidence appears to be that the basic DNA code (the letters C, A, G, T taken three at a time to make syllables) has never changed in the history of the Earth. Nature uses the same code system whether it wishes to make a single cell organism or an elephant. This could suggest that the code is innate in Nature.

## A dramatic statement about the extent of DNA statements

We have noted that the DNA has a molecular weight of some millions and perhaps corresponding to some hundreds of thousands of chains of syllables, of twenty varieties. But from this can be permuted an almost infinite number of words and statements.

*Quoted in H. F. Judson, *The Eighth Day of Creation* (London: Cape, 1979).

Professor Frederick Sanger, double Nobel Laureate, has stated that if some DNA were made the subject of a computer printout it would extend for about ten thousand miles. That would be the equivalent of about 20,000 rather long books in English.

## The problem of organic levels

So far the molecular biologists have only explored the way in which the DNA gives rise to proteins, the body building-bricks, but they have not found how the DNA controls the progressive architecture of the body. They do not know:

How an eye is formed

How specialised tissues are formed

How Nature repairs damage

What controls species characteristics such as appearance and behaviour.

Commenting on this problem, the molecular biologist Sydney Brenner has stated (also quoted by Judson):

. . . . In the next twenty-five years we are going to have to teach biologists another language still. I don't know what it is called yet; nobody knows. But what one is aiming at, I think, is the fundamental problem of the theory of eleborate systems. And here is *the grave problem of levels*; it may be wrong to believe that all the logic is at the molecular level. We may need to get beyond the clock mechanisms . . . we can't understand anything until we understand all the levels of computation that connect the change in the gene with the change in the development, the growth — or the behaviour. That's an immense task.

The italics above are the author's. It is indeed an immense task since the levels of an organism are at least:

1. The creation of proteins — this is now known.

2. The creation of complete cells.

3. The creation of tissues or cell aggregates.

4. The creation of organs.

5. The creation of the whole body. And

6. Characteristic specie behaviour and instincts.

Thus the molecular biologist has only just scratched the surface of the total problem. It is a good beginning to have identified how the building-bricks are created, but the multi-level architecture problem is of a different order.

# 14

## THE SPECIFICITY OF HAEMOGLOBIN

In our anecdotal exploration of monkeys with typewriters we came to the conclusion that there was no chance of any conceivable number of monkeys typing randomly the first verse of Wordsworth's poem *Daffodils*.

The question arises as to whether we can apply similar analyses in the field of molecular biology, particularly in relation to those aspects which the biologists find specific such as DNA or a protein. Our monkey study showed that there is not sufficient time in a conceivable universe for specificity to emerge, and we wonder if this is also true of organic molecules. If this were to be so then it would be a most interesting finding since it would suggest that *life cannot occur by chance*.

### The specificity of haemoglobin

One of the most important proteins, perhaps the most important, is haemoglobin: it is responsible both for the red colour of our blood and for the oxygen chemistry based on our breathing. Fortunately we know a great deal about this molecule.

The formula to describe the specificity of haemoglobin is the same as we used for those typing monkeys and is:

$$\text{Specificity} = \frac{N!}{n_1! \times n_2! \times n_3! \times \text{etc.}}$$

where:
$N$ is the total number of amino-acids in the chain ( = 574 for haemoglobin); and $n_1$ $n_2$ etc. are the numbers of distinct types of amino-acids spread over 20 types.

*!* is the cipher requiring the factorial expansion of numbers such as:

$$1 \times 2 \times 3 \times 4 \times 5 \ldots \ldots \times N$$

The chemical analysis of haemoglobin to enable one to put numbers to $n_1$ $n_2$ etc. is:[*]

| Amino-acid | Number of amino-acids |
|---|---|
| Gly | 36 |
| Ala | 68 |
| Ser | 31 |
| Tyr | 30 |
| Pro | 25 |
| Val | 56 |
| Ile | 1 |
| Leu | 69 |
| Phe | 28 |
| Thr | 14 |
| Try | 4 |
| Cys | 5 |
| Met | 6 |
| Asp | 47 |
| Glu | 29 |
| Amide N (incl. Gln and Asn) | 38 |
| Arg | 12 |
| His | 32 |
| Lys | 43 |

Using the formula just described for alternate linear arrangements of the above amino-acids indicates about $10^{650}$ permutations *only one of which is haemoglobin.* Thus:

[*]Derived from C. U. M. Smith, *Molecular Biology*, p. 109.

Specificity of haemoglobin $= 10^{650}$.

(The actual calculated number was $7.4 \times 10^{654}$.)

## Some large numbers

$10^{650}$ is an enormous number. Compare this with number counts of a cosmic nature:

Number of seconds back from now to start of life on earth (2,500 million years): $10^{17}$

Number of seconds back from now to estimated date of Big Bang $4 \times 10^{17}$ (say $10^{18}$).

Number of atoms in the universe: $10^{80}$.

Number of photons in the universe: $10^{88}$.

Number of stars in the universe: $10^{22}$.

Number of wavelengths of light to traverse the universe: $2 \times 10^{33}$.

## The meaning of 'never' and 'impossible'

Even if the requisite amino-acids had been shuffled once a second through all the time of life on Earth, haemoglobin could 'never' have been produced by random chance. The words 'never' and 'impossible' are difficult concepts, but the physicists have studied such words and have defined 'never' in terms of 'zero in any operational sense of an event'.* This is a correct logical definition since we live in a *finite* universe and so nothing can have happened which would need more time than the life of the universe.

---

*Kittel and Kroemer, *Thermal Physics*, p. 53.

## The specificity of the DNA

Since the DNA codes for *all* the proteins of a creature, it is clear that its specificity must be considerably greater than that just given for haemoglobin above. I do not have an exact figure for the specificity of a given DNA, but one can make a reasonable approximation related to the T4 phage, a tiny creature which preys upon bacteria; its DNA must be one of the smallest specimens. The following calculations hold:

Molecular weight of T4 phage DNA: $1.2 \times 10^8$.

Molecular weight of codon triplet: 1,950.

Therefore number of codon triplets: 61,538 — say 60,000.

Assume that 20 varieties of codon triplets are of equal numerical occurrence.

Therefore there will be about 3,000 codon triplets of each of 20 varieties.

Given the above, one can solve for the specificity of T4 DNA:

Specificity $= 60,000!/(3,000!)^{20} = 10^{78,000}$.

## Darwin's theory of evolution is wrong

Darwin's theory of evolution relies on chance mutations being crystallised by natural selection or the survival of the fittest. But it totally underestimated the time duration which such a theory would need: trillions of times longer than the existence of the universe. In Chapter 10 we studied the problem of dealing with complex permutations and the immense numbers involved in even coming to a chance production of a simple statement such as one verse of Wordsworth's *Daffodils*. We then analysed how Huxley was hopelessly wrong in stating that six monkeys allowed enormous time would randomly type all the books in the British Museum when in fact they could only type half a line of one book if they typed for the duration of the universe.

In this present chapter we have seen how the specificities (improbabilities) involved in organic life are of incredible degree. Haemoglobin has an improbability of $10^{650}$ while the DNA of the T4 bacteriophage has an improbability of $10^{78,000}$.

In a universe only $10^{18}$ seconds old it is obvious that life could not have evolved by chance.

## Postscript

The exact calculated specificity of haemoglobin was $7.4 \times 10^{654}$. There are indications that some of the amino-acid positions are 'neutral' like spaces which are insignificant. The literature indicates that these may be up to 10 per cent of positions. If one accepts these neutral positions, then the calculations can be redone as though haemoglobin consisted of 516 rather than 574 significant amino-acid positions, in which case the specificity reduces to $7.9 \times 10^{503}$. This reduction does not change the tenor of the argument against Darwinian evolution.

In their book *Evolution from Space* Sir Fred Hoyle and Professor N. C. Wickramasinghe use very similar statistical arguments to those we are using here, but they use a simplified formula for calculating biological specificities as $20^N$ where the 20 represents the alternative possible amino acids and $N$ is the number of amino-acids such as haemoglobin in the chain. Their formula is not so accurate as that used by us here, which takes into account the known analysis of haemoglobin as to the proportions present of each. Applying their formula, there is agreement in broad outline as the table overleaf shows:

CALCULATED SPECIFICITY

| | Foster | Hoyle-Wickramasinghe |
|---|---|---|
| Assuming haemoglobin completely specific over 574 amino acids | $7.4 \times 10^{654}$ | $10^{850}$ |
| Assuming 10 per cent neutrality within the specificity i.e. applying to 516 significant amino-acid positions | $7.9 \times 10^{503}$ | $10^{654}$ |

In all cases we are in the realm of 'infinity'.

# 15

## STOCKTAKING: FACTS AND MYSTERIES

### The three facts of specificity

In this Part 3 it is hoped to have established three scientific facts:

1. The basic entities of the universe through the ladder of radiation, atoms and molecules show exact specificity related to numbers. The numbers are associated with cosmic constants, with integral quantum numbers and with the Rule of Eight.

2. As the ladder reaches into the field of biology, the basis of specificity complexifies into 'arrangement' where the numbers are related to permutations of molecules, such permutations forming a coded system.

3. In working out the coded complex of haemoglobin, we found it had the astronomical specificity represented by the number $10^{650}$.

### Emergent form

Overall, the system appears to be one in which numbers as undifferentiated succession, a mere string of numbers as in wave radiation, progressively weave a pattern which becomes manifest as *form* (what I have also referred to as arrangement). This form first manifests itself in atoms where the 'numbers' are electrons and nucleons and the form is that sort of organised quantum arrangement which we might describe as *organism*. The numbers are still there, but there has emerged a formal whole which is greater than the sum of the numerical parts.

But when such organised atoms and molecules enter into biology, the form vanishes as meaningful physical form and re-emerges as *form for form's sake*, i.e. a *linguistic code* or statement.

## *The mystery*

We are alerted to the presence of a mystery by the specificity of protein molecules such as haemoglobin, which we have shown could not have happened by chance. This suggests the existence of some causal world inhabited by 'Maxwell's sorting demons' who can inject unshuffling processes into situations where shuffling might have been considered the dominant influence.

But the analogical move towards 'sorting demons' is paralleled by the indication that the DNA is a coded system. Both indicate some sort of mental background.

Our mystery is:

HOW IS THE DNA PROGRAMMED? IS THERE A 'PROGRAMMER'?

# PART FOUR
# APPROACH TO LOGOS

The DNA is as specific as the proteins such as haemoglobin which it programmes. But what is the origin of its specificity? To name the unknown, let us call it LOGOS.

In considering the nature of cosmic programming we move into the realms of software, and in this Part 4 we consider some of the relevant aspects of software programming as understood from our man-made technologies.

# 16

## APPROACH TO COSMIC SOFTWARE

In searching for 'what is behind the DNA' it would seem that we enter the realm of *software*. Molecular biology can find no trace of further hardware which is upstream from the DNA, and since the DNA is known to be coded, then we are not looking for more physical facts but for mental functions.

Until the invention of electronic computers such an approach might have been considered as pure metaphysics, but the opening up of the computer art tells us that software is both 'real' and as important as hardware.

### Invisible data — 'white noise'

If we could reduce our physical size by taking some of Alice's 'Drink Me' so that we could crawl about the insides of an electronic computer, we would find that there was almost nothing there. There would certainly be nothing worth seeing. Even if we could tap in to the main electrical systems and attach an amplifier and a pair of headphones, all that we would hear is a noise like escaping steam as millions of invisible electrical digits rushed along in the time-stream. Because the digits are random, except to the programmer, the noise is 'white noise' without any distinguishing characteristics. Yet that white noise may be controlling the navigation of an aircraft or doing the payroll of a factory.

Even if we could slow down those digits by a factor of a million or so, all we would hear on listening-in would be something very like the Morse Code. The reason why we cannot find any human 'meaning' inside a computer is because there *is* no human meaning inside a computer: all the human meaning is either at the input in the mind of the programmer or in the output where the programmer can look for ultimate results. Between the input and output of the computer, data is being processed, and the particular method used to organise the data processing is what we now call software, the rules for digital thinking. These rules are *pure logic*.

## 'Logos'

If we now transfer our thoughts from man-made computers to 'what is behind DNA', we have little choice but to imagine that there is a correspondence. Now 'what is behind man-made computers' is not a thing; it is pure logic. In the DNA we have seen the 'thing' or hardware of natural computing, but we need to invent a term for the logic of the system and there seems no more appropriate word than LOGOS. This Greek word means word or reason, the mind-stuff itself. The word also has religious connotations which are not intended at this stage, although later we may find them suggestive.

## Surveying 'Logos'

In approaching the subject of LOGOS we need first to establish a number of ideas which may be relevant in this somewhat abstract field. In particular:

1. *The nature of language.* We need to be quite sure that language is a technique for coded specificity, and in this we shall keep very close to the nature of Queen's English. This is an anthropomorphic approach, but we have no other. The hope is that study of our native language could provide an analogue as to the literacy of LOGOS. (Chapter 17)

2. *The nature of control.* Here we are concerned with how coded specificity in language can manifest as control specificity in the physical world. For this we need to study a little cybernetics, which is concerned with such processes in man-made systems. The DNA is not only coded information; it is also executive in control of organic chemistry and structure (Chapter 18).

3. *Programming privacy.* From the above we shall find that mind is innately private and that control in nature is a private-public transformation. This is basic and gives the reason why one can never 'see' LOGOS (Chapter 19).

4. *The Extended dimensional continuum.* In order to accommodate the private-public data transformation, an extended dimensional system is needed more than that provided by conventional Space-Time. To this must be added Imaginary Space-Time to form a continuum with it (Chapter 20).

5. Since it is obvious fact that the Sun is the supporter of all organic life and creates the terrestial conditions appropriate for life, we make a brief study of *Solar Suggestibilities* (Chapter 21).

Given the possible relevance of the above five facets, we explore the nature of LOGOS further in the final Part 5.

# 17

## THE NATURE OF LANGUAGE

We now examine the phenomenon of human language to find whether it throws light, even by analogy, on the high degree of specificity to be found in biology and living creatures. Human language appears to be a coded system for expressing ideas at increasing levels of specificity whose ultimate is Truth.

Consider the following alphabetical characters:

AABCCEHIKLSTT

As thus stated the string is meaningless, and all we can recognise is that there are three pairs of letters (A, C and T) which are the same. The mathematics of possible arrangement of these thirteen letters is in 778,377,600* ways, and so we can state that they represent a system of high entropy and low specificity.

But now a linguist comes along and shows that these letters can be arranged in a more meaningful pattern:

BLACK THE IS CAT

It is not a very meaningful statement, but we begin to recognise the appearance of words. The alphabetical letters have been partly unshuffled or sorted. So what we now have is a system of medium entropy and medium specificity. But now we have a Eureka moment and decide that the correct arrangement of the words should be

THE CAT IS BLACK

* $\dfrac{13!}{(2!)^3}$

So now we have moved into a situation of low entropy and high specificity. Out of those 13 originally jumbled letters we now have a simple and specific statement of meaning which we can all understand.

## The nature of coded statements

Because this book is written in English for English-speaking people, it may not be obvious that all the above statements from AABCCEHIKLSTT to THE CAT IS BLACK relate to *code*, but this would be fairly clear to a Chinaman or in the case of an equivalent statement in Chinese to an Englishman. Because of inheritance and education we take for granted the fact of living in a world of codes. But what is a code?

A code is any word which we agree shall correspond to an element of experience of either physical or mental nature.

But note that the original letters of the alphabet are themselves not really a code; they are *different symbols*, and it is by arranging different symbols in agreed strings that we create word codes which are agreed to correspond to elements of experience. The creation of a code system which is to be effective involves a series of hierarchical stages or levels:

1. A set of different marks which can be letters, numbers or other symbols whose main property is that they are *different*. One could use apples and oranges. The simplest of all mark-differences is that in which one can only distinguish 'this from that', which has led to the Morse Code (dot and dash) and the binary system used in computers (0 and 1). Alphabetical marks using 26 letters are a very complicated basis for a code system, but this is due to the extra need to cater for phonetic syllables in human speech. However, we will by-pass that diversion.

2. The different marks are arranged in an agreed order to corres-

pond to words as elements of experience. Note the importance of the *agreement* since we are not aware of any absolute coding system which does not require agreement as to meanings. This brings a coding system into the phase of words and language.

3. The coded words themselves need to be brought into a limited set of permissible relative arrangements, and we agree this as *grammar*. Thus of the two statements given earlier one is non-grammatical while the other is grammatical:

BLACK THE IS CAT — non-grammatical.
THE CAT IS BLACK — grammatical.

4. Given words in correct grammatical arrangement, they must agree with established systems of logic. Thus the two following statements are both grammatical but only one is true (they use exactly the same words):

ALL ANIMALS ARE CATS — false.
ALL CATS ARE ANIMALS — true.

Thus at the very least an effective coding system depends upon:

1. Different marks ('letters') whose relative arrangement creates words.

2. Agreement that the words correspond to a public meaning as to an element of experience.

3. Word arrangements must conform to agreed rules of grammar.

4. Word arrangements must conform to agreed rules of logic.

## Language can create high specificity

A coded statement, for example in English, can lead to any degree of complexity simply by increasing the length of the statement to cater for its conditional logical restrictions. Compare the following two statements:

I WILL PLAY GOLF TOMORROW

I WILL PLAY GOLF TOMORROW IF IT DOES NOT RAIN AND THE CAR IS AVAILABLE AND GEORGE CAN ALSO PLAY.

The first is a simple statement of high specificity since the only condition is TOMORROW which itself is specific. The second is also a statement of high specificity in laying out more conditions, but it is more complex than the first statement.

Thus statements of high specificity can vary in terms of relative complexity, and language copes with increase of complexity by increasing the length of its statements. There is no limit to this.

## Language and statements of low specificity

A language statement has low specificity when it can be interpreted in alternative ways. A classic example was written by Lewis Carroll:

> *Twas brillig and the slythy toves*
> *Did gyre and gimble in the wabe.*

This is a statement using correct grammar but containing six words (brillig, slythy, toves, gyre, gimble, wabe) whose meanings are unclear and therefore subject either to alternative interpretation according to individual fancy or even to no interpretation at all. It is a statement of high entropy and low specificity. This shows that language is only meaningful when all the word meanings are agreed. It implies that somewhere in the background there is a *dictionary*, actual or notional.

This enables us to make further definition of a code system:

A code system is one in which arrangements of symbols into words have agreed meanings registered in a dictionary (the 'code book').

## The purpose of language

What is the purpose of language? It is by no means certain that language would be essential to a human being on a desert island. I don't really need to utter the words THE CAT IS BLACK in order to understand that the cat walking in front of me is black. But as soon as I am involved with other human beings, then language proves to be an invaluable means for communication. So far we have mainly considered language as an ingenious system for making specific statements based on an arrangement of marks as codes (in speech the 'marks' are acoustic syllabic phonetics). But in developing the procedure we noted that there had to be *agreement* as to what the codes meant. The code-word itself without a notional code-book is meaningless, and it is the use of the notional code-book which converts a word from high entropy to high specificity.

## Human communication using language

Communication by language is quite complex and consists of five sequential steps:

1. I (as Mr A) have had repeated experiences of actual black cats, and such experiences belong to my imagery or imagination.

2. I wish to communicate the notion of a black cat to Mr B, so I look up my notional visual image code-book and find that black cat is represented by the code BLACK CAT.

3. But the figurative code words BLACK CAT also have an acoustic

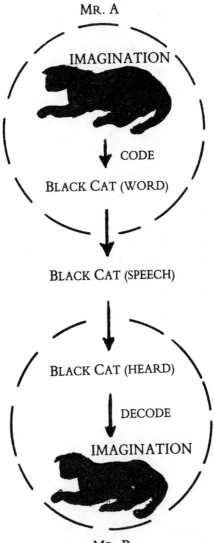

Fig. 17.1. Communication: imagination to imagination.

translation into the spoken word BLACK CAT as a series of phonetics.

4. So I do the phonetic translation and utter the words BLACK CAT so that Mr B can hear them.

5. Mr B, receiving this phonetic code, must refer the received sound to his phonetic-visual code book in which BLACK followed by CAT, as dictionary words, are accompanied by a picture of a black cat. This picture is in Mr B's imagination and he therefore recognises my message to him. We are now on the same wavelength.

But note the process:

1. Mr A's imagination
2. Image-word-phonetic coding
3. Speech
4. Phonetic-word-image decoding
5. Mr B's imagination.

Thus the process is one where the content of Mr A's imagination is transferred to Mr B's imagination via the language link and as drawn out in Fig. 17.1. We see the need for language since there is no known means whereby Mr B can have direct knowledge of what is going on in Mr A's imagination. So we can define language:

Language is a system whereby two *invisible* imaginations can communicate with each other through a word-coding system.

We shall later consider (Chapter 19) just why it is necessary for individual imagination to be invisible and intangible.

## The second use of language systems — control

So far we have only considered coded language systems as a means for *communication* between two human beings (no doubt bird-song is an equivalent means for communication between birds). But there is a second quite distinct use of language, and that is for the purposes of physical creativity whereby an idea becomes a physical manifestation. This now takes us to control theory (cybernetics).

# 18

## CONTROL THEORY (CYBERNETICS)

If we stand back from reality, we note that it consists of 'systems'. The systems may be immense such as a galaxy; at a smaller level one has a system such as the solar system. Further down the scale one may have a system such as the Earth with its prime components of core, lithosphere, hydrosphere, atmosphere and biosphere. Within the biosphere we come to a smaller system, that of an organic creature such as man. Further down we may well consider that molecules and atoms are also systems.

The point about a system is that it is a whole made up of parts and with the likelihood that the whole is more than the sum of the parts. But the question is how a system is controlled so that the parts co-ordinate together to create a harmonious whole.

The science of such control is called cybernetics. This name was coined by Dr Norbert Wiener, and when I once asked him what it meant he told me that it was the Greek for 'the art of the steersman'. So cybernetics is concerned with the steersmanship of systems for their benefit.

### The two control objectives

Any system can have two objectives:

1. To maintain its identity, to continue to be such a system and not to fall apart. This objective could be called *conservation*, and the cybernetic method whereby it is achieved is *negative feedback control*.

2. To *progress*, to evolve into a more favourable or different state even though this means leaving behind its former identity; the cybernetic method whereby this is achieved is *navigation control*.

## Conservation through negative feedback control

The purpose of negative feedback control is to ensure that a system is kept in, or restored to, some ideal state. The simplest example is that of a house thermostat which can be set to control the heat supply to keep the house at some ideal temperature such as 68°F. The schematic for negative feedback control is shown in Fig. 18.1. Let us adhere to the house thermostat as the example:

1. There is a device which declares the *ideal state* of the system (for example 68°F).

2. There is a device which is measuring the *actual state* of the system. It may be a thermometer which is registering 64°F, and thus the house is too cold by the ideal standard.

3. The ideal state and the actual state are compared in an *error comparator*, which declares 'The house is 4 degrees too cold'. This is the error signal.

4. The error signal goes to the *actuator* which has two programmes:

(*a*) To turn up the heat if the house is too cold.

(*b*) To turn down the heat if the house is too hot.

In the instance given of the house being too cold, the actuator will turn the heat up. This is *correction behaviour*.

In this fashion the house will warm up until it may somewhat overshoot the ideal temperature and then an error signal will be generated in the opposite sense to turn the heat down.

Such a system is called 'negative feedback' because the correction behaviour is in the opposite direction from the error signal, i.e. the correction behaviour *negates* the error signal.

This type of system is therefore appropriate to keep a process more or less close to its ideal state, and it is the major control law used by nature for organic creatures. Most typical in nature is

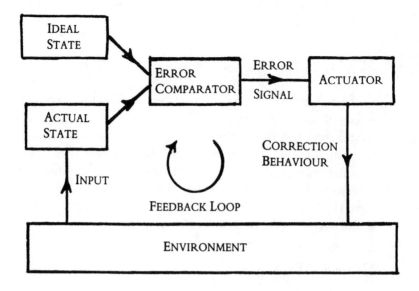

Fig. 18.1. Negative feedback control for stability.

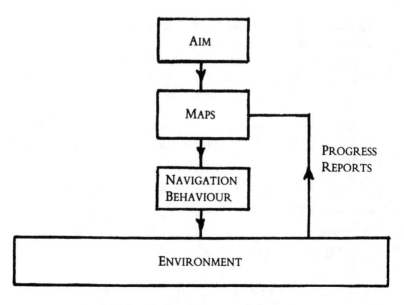

Fig. 18.2. Navigation control for progress.

where the Error Signal is 'hunger' and the correction behaviour is 'foraging for food'.

## Negative feedback control and behaviour specificity

In the organic world the control systems are mainly devoted to 'survival' and the life-wish. Without the system described a creature could be in one of many states of ill-health or disablement and thus in a state of high entropy and low specificity. But bringing the creature under negative feedback control means that it has a single aim (survival) and so all its actions are devoted to this end. Thus its behaviour becomes specific, a *single aim*. It could alternatively be described as 'single-mindedness'.

## Navigation control

The complementary control system to the above is one which can be defined:

Navigation control is that control arrangement which enables a system to achieve an aim or improve its position beyond that of mere stability or conservation. Thus the 'ideal state' is a novel state which has not been previously experienced. The control might be described as relating to a climbing process.

A typical example is that of a ship making a voyage from Port A to Port B. The control arrangements are as shown in Fig. 18.2 and:

1. There has to be a declared *aim* for the process, and in the nautical example it would be 'Port B'.

2. There has to be a *map* or model of data which shows the zone to be navigated and on which is marked the best route (say to get from A to B).

3. There have to be *navigation skills* whereby the aim can be realised with the aid of the maps.

## Navigation control and behavioural specificity

The achievement towards a specific aim channels the behaviour into a single line of action, and this reduces the entropy of the system by eliminating alternative courses and establishing a state of single-mindedness.

## Double control systems — DNA

Many systems operate simultaneously under negative feedback and navigation control, the former providing the stable background as a basis for the excursions of the latter. So we human beings eat and drink under negative feedback control in order to preserve our health, while our external behaviour may be some form of work under navigation control. It would seem that the DNA is a dual control system of this nature since it has functions relating both to the stability of the body and the variations of external behaviour. The two aspects can also work together for the same purpose, as when hunger (an error signal under negative feedback control) causes an animal to forage for food under navigation control.

## Higher and lower languages in control systems

Any control system is essentially a data processing system which has different levels. The different levels are necessary in order that the integrity of the aim should be protected from the behavioural aspects of the system, i.e. the behaviour must not be allowed to change the aim. This applies equally to negative feedback and navigation systems, and it is achieved by a *change in language* between levels and with a coded interface between levels.

Consider the case of the ship navigation from A to B:

| Aspect | Type of language |
|--------|------------------|
| Aim | A few words such as 'New York' (as Port B) |
| Maps | Graphical (shapes) maps |
| Behaviour | Physical: motion to wheel or engine controls. |

What we note is that the higher and controlling levels of language are *barely in space-time at all.* It is as though the controlling factors come from a *different non-physical dimension* (later we shall identify this mysterious dimension as Imaginary Space-Time). Even the navigation maps on which the ship's voyage is charted are hardly in physical space-time.

## The master-slave relationship

The relationship as between a programmer and a controlled process is a master-slave relationship in which the slave does what he is commanded by the master to do. So the direction of command must be one-way from master to slave and not from slave to master. This is ensured by the change of language in the chain of command, as we next consider.

# 19

## PROGRAMMING PRIVACY

Let us consider two main ideas which have been developed in the last two chapters:

1. In Chapter 17, on the specificity of human language, we concluded that a main purpose of language is to provide a communication link as between two *invisible* imaginations. This is empirical experience and fact, and it is remarkable that my imagination is *private*.

2. In Chapter 18, on control theory, we noted that control is a master-slave relationship and that the programming intent of the master is hidden from the slave. This secrecy is due to the language barriers in the chain of command, but the barriers are one-way only. This is because the higher or master level is *imaginative*, and imagination can understand lower levels such as maps or physical skills, but nothing below the level of imagination can understand imagination.

So we note that in both cases integrity, whether of communication or control, is maintained by the privacy of imagination, by the imagination keeping a due distance from the physical world.

### Programming privacy

By integrity I imply specificity, the importance of the fact that the master controls the slave. Almost all physical processes are wayward and subjected to shuffling and disorder, and they can only be controlled by a programme which is specific. This raises the physical process to a corresponding specificity. But if the programme is to be specific, then it has to be internally de-bugged; that is, it has to shed all possible alternative configurations at the moment of action. It has to be decisive in a categorical fashion.

Now, since programmes originate in imagination, all alternative configurations and second thoughts have to be put aside *in private*. Note that imagination itself is not specific unless it decides to be specific, for imagination can dream all over the place. So the imagination has to sort itself out in private, otherwise it cannot crystallise a programme for external specificity. Note that this discipline applies not only to the imagination itself but also to the maps and models it employs. So we must consider maps and models as also private and not to be deployed until correct courses have been plotted on them. This is certainly true of the human being — that all its inner data are private.

## The privacy of 'Logos'

If we regard LOGOS as the programmer of the DNA, then these arguments suggest that LOGOS must be invisible and always so. Indeed, any possibility of a human being having direct contact with LOGOS would need to be through the imagination.

# 20

## DIMENSIONS FOR LOGOS

In considering the transmission from the world of ideas to the world of physical fact, there appears to be a bridge between the two — a sort of ladder of language with three rungs:

1. higher language, based in the world of ideas and imagination;

2. median language, which can code a specific idea into words, mathematics or blueprints; and

3. lower language, which interfaces with physics and chemistry, the lower language itself being a construct of physics and chemistry.

In this three-rung ladder the central key transition is a *code*, for it is coding that which is the critical interface between idea and fact, between imagination and 'thing'. Thus the code links upwards to ideas and imagination and downwards to physical reality.

### *The human being as the typical embodiment*

There is no better example of the embodiment of the above situation than in the human being where the three rungs are most generally

1. the inner world of mind and imagination;

2. the translation of the above into the formality of words, mathematics or blueprints as 'the halfway house' symbolic coding; and

3. the level of physical behaviour in speech and action.

## The need for a dimensional system

In our ordinary affairs we find it useful to have a notion of a background of dimensions such as space and time as a sort of blank canvas for our concepts and constructs. The notion of space is obligatory to provide a canvas for the world of form 'out there', and the notion of time is obligatory to provide a canvas for the movement of form inherent in our experience of past-now-future. But it is not two canvases but a single canvas with two aspects, an aspect of structure and an aspect of change: it may be called Panoramic Space-Time.

## Panoramic Space-Time: pST

The word panorama comes from Greek and means 'comprehensive view', and thus Panoramic Space-Time is the canvas for our view of the world 'out there'. For short I shall call this pST. I also use this expression involving 'panorama' because it coincides with relativistic concepts of space-time which require an observer to validate events. I think this an important factor for we cannot conceive of a space-time which is a vacuum without associated events and an observer for such events.

## Imaginary Space-Time: iST

But we have experience not only of Panoramic Space-Time 'out there'; we also have an equivalent experience of a sort of inner panorama 'in here' which one may call Imaginary Space-Time or iST. The inner panorama was well expressed by the philosopher David Hume (1711–76) in his *Treatise of Human Nature*:

> For my part, when I enter most intimately into what I call myself, I always stumble on some particular perception or other, of heat or cold, light or shade, love or hatred, pain or pleasure. I never catch *myself* at any time without a perception,

and never can observe anything but the perception. . . . There may be some philosophers who can perceive their selves but setting aside some metaphysicians of this kind, I may venture to affirm of the rest of mankind, that they are nothing but a bundle or collection of different perceptions, which succeed each other with inconceivable rapidity, and are in a perpetual flux and movement.

Now what Hume describes is a mixture of phenomena coming from pST and reflected in iST. For example, the very idea of a percept is that it is both something coming into one from the outside and yet involves a recognition faculty which is inside the person. But the question as to whether we have experiences which are purely in iST is simply answered, for these we know as *dreams*. When I dream at night I am cut off from percepts coming from pST, for in dreams there is no real external panorama. What I experience is an inner panorama in iST.

Thus I conclude that the human being is affected by two sorts of panorama, one 'out there' in pST and one 'in here' in iST.

## The pST-iST Continuum

If we can agree that we human beings are subjected to two sorts of panorama, one of which relates to 'out there' and one of which relates to 'in here', then the question is what is their relationship. But note that they are distinct:

*Example of Independent pST and iST.* I am driving a car from A to B so there is no doubt I am involved in pST 'out there'. But at the same time I am worried about my bank account and am wondering how to improve my money affairs. This has nothing to do with driving the car and takes place in iST 'in here'.

This may serve as an extreme case of split activity taking part in two sorts of world. But next consider:

*Example of Co-incident pST and iST.* I am reading a very interesting book. There is no doubt that the book is 'a thing out there' and so in pST. But the reading of the words is triggering all sorts of inner thoughts and images and associations, and this is taking place 'in here' in iST.

So we find that our mind can co-incide with the external panorama or it can be independent of it. As the cyberneticians would say, we can be 'on-line' or 'off-line' to reality. The question arises as to whether we can find a symbolic representation which holds good for both cases. Fortunately we can, and it was invented by electrical engineers whose currents and voltages are sometimes co-incidental (in-phase) and sometimes independent (out-of-phase). The electrical engineers use the terms *real* and *imaginary* for the two aspects and have a symbol '$i$' for manipulating the mathematics of the imaginary ($i = \sqrt{-1}$).

The dimensional diagram we need is shown in Fig. 20.1 and:

1. There is a horizontal axis of Panoramic Space-Time (pST) to accomodate events 'out there'.

2. There is a vertical axis of Imaginary Space-Time (iST) to allow for events 'in here'.

3. The area included between the two axes is a Continuum (pST-iST Continuum).

4. Real events consist of Reality Vectors into the Continuum such as are represented by $A$ or $B$ or $C$ and which have different angles corresponding to different relative amounts of 'out there' and 'in here'.

Note how this construction can cater for the two examples given (Fig. 20.1). Thus vector $C$ is the driving of the car almost entirely in pST while vector $A$ is the thinking about financial worries in iST. Vector $B$ is the reading of the interesting book in which there are about equal contributions from pST and iST, and the book as 'thing' and the book as interesting meanings co-incide.

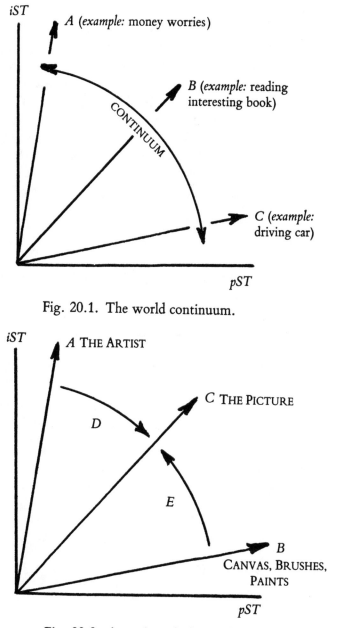

Fig. 20.1. The world continuum.

Fig. 20.2. An artist painting a picture.

## The nature of reality

Reality is an experience *now*. At one end of the Continuum spectrum it could be 'looking at a house, now' dominated by pST. At the other end it could be thinking about what happened to Julius Caesar 2,000 years ago but noting that such musing is also *now* in the iST of imagination. Thus one does not discriminate as between reality 'out there' or 'in here' since either can be valid reality if it relates to experience *now*. This erects the relativity observer to an eminent position since only an observer can state 'now' and keep on stating it with the lapse of time.

## Further examples of the Continuum

The following are some principal states of the Continuum:

THE ARTIST OR CREATOR

Consider an artist painting a picture.

1. His concept of the picture is in his imagination in iST at the vector $A$ in Fig. 20.2.

2. The materials he will use such as canvas, brushes and paints are almost entirely 'things' in pST at the vector $B$.

3. The actual picture emerges as the imagination and the materials come together as the vector $C$. Thus the original vectors $A$ and $B$ come together (see arrows $D$ and $E$) to merge in vector $C$.

HUMAN COMMUNICATIONS

Consider Mr A talking to Mr B (as in Chapter 17):

1. What Mr A wishes to convey is private in Mr A's imagination in his iST 'in here' as shown in Fig. 20.3.

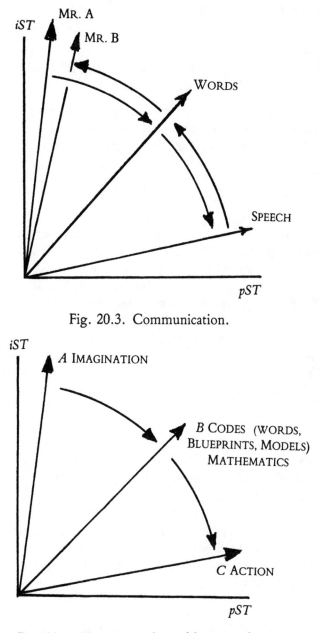

Fig. 20.3. Communication.

Fig. 20.4. Human psyche and language levels.

2. Mr A turns his imaging into the coding of words which occupy the middle ground of the Continuum since they are silent 'data things'.

3. Mr A speaks out the words as action-speech, which is almost entirely in pST as an acoustic wave.

4. From this public speech the hearing of Mr B extracts the words and refers them to his code-book (memory dictionary), and then the meaning goes to Mr B's imagination in iST.

## Dimensions and human beings

The 'mystery' we are pursuing is 'what is behind, or programming, the DNA', and we now examine our dimensional construction to see 'where' LOGOS is, LOGOS having been defined as the mysterious programmer. Referring to Fig. 20.4, we first establish the general construction for a human being:

1. The higher language of the human being is essentially imaginative and must be almost entirely in iST as shown by the vector $A$. This fact has been very clearly described by Jacques Monod, the molecular biologist and Nobel prize-winner:*

> I am sure every scientist must have noticed how his mental reflection, at the deeper level, is not verbal: it is an *imagined experience*, simulated with the aid of forms, of forces, of interactions which together barely compose an 'image' in the visual sense of the term. I have even found myself, after lengthy concentration on the imagined experience to the exclusion of everything else, identifying with a molecule or protein. . . . It is the powerful development and intensive use of the simulating function that, in my view, characterize the unique properties of man's brain.

---

* Jacques Monod, *Chance and Necessity* (Eng. transl. London: Collins, 1971, pp. 145–6).

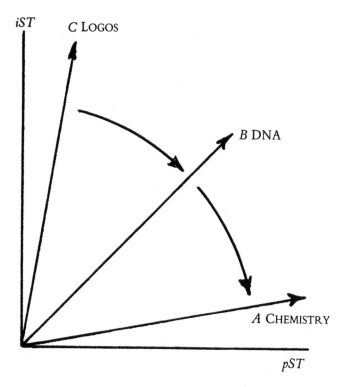

Fig. 20.5. Organic creation and languages.

This description exactly fits with my definition of a code (Chapter 17), namely that a code is any word which corresponds to an element of experience. Monod is suggesting, and I agree, that it is imagination which is 'behind' experience or code.

2. Imagination in its progression of approach to physical actuality has to adopt a coded form which embraces both its origins of imagination and its ultimate contact with the physical world. This median language is that of words, blueprints or models, a sort of semi-crystallisation (in biology it is the DNA). This is shown in Fig. 20.4 as the vector *B* having about equal projections in iST and pST. In the world of human beings the most common manifestation may be 'conversation'.

3. The third degree of lower language in human affairs is action as

shown by the vector *C* of Fig. 20.4, and is almost entirely in pST for it is 'visible and tangible' to our friends and acquaintances. It is the factual world of creativity under navigation control.

In this way the inner invisible private world of imagination is ultimately manifested as the outer public world of action and creativity.

## Dimensions and 'Logos'

If we, in anthropomorphic analogy, apply the above analysis to LOGOS, it is possible that an equivalent state of affairs exists. All we have defined for LOGOS is that it is a 'name' to represent 'that which is behind the DNA' — a sort of '*X*'. Dealing with the dimensions for LOGOS, it is logically simpler to take the factors in reverse order as shown in Fig. 20.5:

1.  There is a lower language which, as the vector *A* essentially in the dimension pST, is chemistry. Everything here is physical and can be understood in terms of what we can or might see under the microscope. It is public, and protein is typical.

2.  The median language represented by the vector *B* is the DNA which is a projection partly in pST and partly in iST and can only be half understood. It can be understood under microscopes as a molecule, but its other half as a code cannot be understood. Thus the DNA is the interface between understanding and mystery.

3.  'Beyond the DNA' it can be suggested that we are in the realms of the vector *C* in iST, the world of causal imagination and thus of LOGOS itself.

# 21

## SOLAR SUGGESTIBILITIES

In considering factors which might be of significance in answering the question 'what is behind the DNA?' and which we have tentatively named as LOGOS as a mysterious Factor X which may or may not exist, it will be obvious that somewhere the Sun as the major obvious source of life comes into the equation. In this chapter we briefly review those factors favourable to organic life on Earth which obviously emanate from the Sun. It is a truism that if the Sun went out then all life would vanish.

### *The Sun provides the energy for photosynthesis — Factor A*

As we saw in Chapter 13, the solar radiation provides photons which, falling on green vegetation, raise an electrical potential within plants as their prime source of energy. It is this energy which enables the two main plant foods of water and carbon dioxide, which are otherwise in a ground energy state, to raise their vitality for chemical conversion to carbohydrates and sugars. These the plant can burn at its convenience to suit its various energetical needs.

### *The Sun provides a thermostatically-controlled climate — Factor B*

All chemical processes, including bio-chemical processes, are subject to variations with temperature. So a degree of thermostatic control is desirable for consistent chemical reactions. In the case of the human body the temperature is controlled within fine limits as to 98°F. plus or minus one degree. The Sun provides such a thermostatic umbrella over most of the Earth through the simple means of evaporating water from the surface of the oceans and dropping this as rain over dry land. This keeps the dry land temperatures down since the Sun has to provide the heat (the latent heat) to evaporate the oceanic water.

In a similar fashion the Sun prevents freezing by melting snows and ice. Overall, about half the dry land surface of the Earth is easily habitable under general thermostatically controlled conditions.

## The dual environmental conditions provided by the Sun — Factor C

The conditions for the survival of organic life are provided by two distinct environments: the oceans and dry land. This is a very important duality, for the oceans excel in the *origins* of organic life while dry land excels in the *evolution* of life. The question occurs whether it is solar influence which has demarcated the surface of the Earth into two distinct regions. While it is easy to account for the solid aspects of the Earth which we share with the other planets, it is not at all clear that oceanic water had solar origins. As we cannot account for water on the surface of other planets (where it is absent), we appear not to be able to account for its solitary appearance as the oceans of Earth. While water appears to be a necessity for life, we must discount its appearance on Earth from solar influences and thus consign Factor C to the 'not-known' list.

## The solar production of soil — Factor D

A known requirement for life on dry land is soil. Without soil, water cannot percolate around the roots of plants, nor can plant roots penetrate to reach water (and other minerals). What is at work which is favourable for life is the break-up of rocks into granules which one day will furnish the basic matrix of soil. This weathering takes place due to a number of mechanisms all of which relate to solar forces. The daily heating and nightly cooling of rocks ensures that their surfaces will develop cracks. The addition of rain turning into ice ensures that the expansive force of ice will widen and breach the cracks. Thus over aeons of time the surface of terrestial rocks is turned into sand and soil.

## *The releases of vital trace minerals — Factor E*

Organic life depends not only upon the major atomic chemicals of hydrogen, oxygen, carbon and nitrogen but on three other central minerals — namely magnesium, iron and phosphorus, all of which are originally rock constituents. Magnesium is needed as the central atom of chlorophyll for plant photosynthesis, iron is needed for the central atom of haemoglobin the blood protein, and phosphorus is needed both for the ATP involved in photosynthesis and for the structure of nervous systems. These three vital trace minerals are washed by rain from the rocks of the Earth and appear both in drinking water and in the oceans.

## *Tides and the littoral — Factor F*

Organic life certainly started in the oceans and partly moved on to dry land. This was possible because of the existence of littorals or beaches where ocean met land, and so the conditions were established for the existence of creatures which could survive in either medium and so gravitate according to the medium of choice. But the littorals were only possible because of the tides which were in turn due to the combined gravitational attraction of Moon and Sun, the tides being at the maximum (and the littorals more expansive) when Moon and Sun were in conjugation.

## *The influence of winds — Factors G*

Winds are a very important condition for the sustaining of organic life since they are the mechanism which ensures that the thermostatic rain system (see Factor B) operates by the motion of clouds from above the oceans to above dry land. Otherwise rains would rise from the oceans and drop again on to oceans.

## Atmospheric protection against meteorites and harmful radiation — Factor H

Quite apart from the need of life to breathe atmosphere, life on Earth would be impossible without the protection of the atmosphere from hostile influences, particularly

1. the function of the atmosphere to burn up meteorites before they can strike the Earth; and

2. the function of the atmosphere, and especially the upper ozone layer, to absorb ultra-violet and X-ray solar radiation.

One only needs a glance at the pock-marked photographs of the Moon to realise the havoc caused by meteorites, and more subtle scientific knowledge convinces us of an equally devastating threat from harmful radiations. But we cannot be sure that these two protective measures are due to solar influence, so this Factor H must be placed in the 'not known' category.

## The total specificity of solar influence

That solar influence is specific in favouring life on Earth might be guessed from the fact that our modern knowledge of other planets, considerably increased since space exploration, suggests that Earth alone might support life as we know it. From our examination above we have found six solar factors which are favourable to life on Earth:

Factor A. Energy for photosynthesis

Factor B. Thermostatic control

Factor D. Soil production by weathering

Factor E. Leaching of vital trace elements

Factor F. Tides and the littorals

Factor G. Winds.

There were also two further factors favourable to life but which are not obviously traceable to solar influence:

Factor C. The division into oceans and dry land

Factor H. Protection from meteorites and radiations.

So let us be satisfied that the Sun provides six major influences favourable to organic life on Earth. Each of these is relatively independent from the other, and so the specificity is

$$6 \times 5 \times 4 \times 3 \times 2 = 720.$$

This is not a tremendous specificity, such as we found for haemoglobin, but it is significant.

## Degrees of specificity

The proof given in Chapter 14 as to the astronomical specificity of haemoglobin, as represented by a figure of $10^{650}$, may have blunted our sense of proportion as to relative specificities. Thus the fact that the solar specificity for life on Earth is 'only' a 720:1 odds-on value may strike us as not very impressive. But if any one of these conditions were missing, then life on Earth as we know it would be impossible.

The fact is that 'improbability' (or specificity) starts when the odds-against exceed 2:1 (heads or tails). Thus a specificity of 720:1 is a very high improbability of accidental occurrence. I conclude that the solar influence in encouraging the environmental conditions on Earth for the growth and maintenance of life is *specific*. We stay in the realms of mathematical science.

# PART FIVE

# WHERE ANGELS FEAR TO TREAD

In Part 4 I have described a number of factors which may be relevant to our quest as to the question 'What is behind the DNA?' The factors were

1. That in considering LOGOS we are in the territory of software and way beyond 'physics'.

2. That the study of our own native language may be the best guide to what we may mean by software.

3. That somewhere cybernetics (control theory) may come into the situation.

4. That programming is always taciturn and private.

5. That (as seems certain) Imaginary Space-Time is a legitimate dimensional system and that it may be a continuum with Panoramic (i.e. physical) Space-Time.

6. That the Sun is highly involved in the drama.

The above are six jig-saw pieces of which some may legitimately belong to our puzzle and some may not. At this stage I do not attempt to present any sort of credible system. So we now move into territory 'where Angels fear to tread'.

# 22

## THE PURPOSE OF LIFE

We have defined LOGOS as 'that which is behind the DNA'. But since we have found that the DNA has a specificity easily exceeding $10^{650}$, it is doubtful if we can ask any direct questions about LOGOS any more than we can ask direct questions about magic or miracles. So what we need to know in the first instance is something *rational* about LOGOS, perhaps not so much about how it works as what could be its *purpose*. If one were faced with an aircraft in a hangar it might be very difficult to diagnose what it is; but if one found that its purpose is *to fly*, then its details of construction might start to make sense. Fortunately, although it may involve some common-sense guessing, it may not be very difficult to discover a main purpose of LOGOS: that could be *to prevent the universe running down to extinction*.

If we human beings and all life of other forms have one main single desire it is the Life-Wish, the wish not to be extinguished, not to die, and our first 'good-guess' is that this may also apply to the universe. For it is difficult to conceive that a universe such as ours, with all its energy and richness of variety, could be willing to allow it all to fade and vanish.

Yet the Second Law of Thermodynamics tells us that, failing some sort of rescue operation, all is indeed on its way to universal death. It was Eddington who said (Chapter 7):

> If your theory is found to be against the Second Law of Thermodynamics I can give you no hope: there is nothing for it but to collapse in deepest humiliation.

So our first good-guess is that LOGOS has, somehow, found an effective way of circumnavigating the Second Law of Thermodynamics.

## The time-scale of 'Logos'

The reader may well wonder why I have chosen the problem of the running down of the universe to start an investigation of LOGOS. After all, the basic question was simply 'What is behind the DNA?', so why bring the running down of the universe into it? My reasons for this are:

1. The DNA code appears to be as old as life, not less than 2,000 million years antiquity.

2. Man himself has only been around for some three million years, which is negligible compared to the lifetime of organic life.

3. 2,000 million years ago might well have been a crisis point in the universe when it was necessary to deal with the otherwise inevitable consequences of the Second Law of Thermodynamics towards the death of the universe. The 'timing' is very suggestive.

So one thing is certain: that LOGOS is very ancient and certainly as old as life, perhaps as old as the universe, and that it applies to all life, for all life has nuclear DNA.

Let us examine the matter under the assumption that Creation took part in two stages, the first stage concerned with matter in gross aggregates, such as stars, and a second stage involving organic life.

No conceivable system can continue to create energy for infinite time. During the First Stage of Creation there may well have been almost infinite energy to start with (the Big Bang?), but with the remorseless passage of time the quality or temperature of the energy must run down under the Second Law of Thermodynamics, so that the universe was on its path to maximum entropy and disorder. What this involved was a progressive reduction in the thermal potential in the universe towards an ultimate point at which all would be at the same tepid temperature and the stars or suns would go out. It would be a dead world and for practical purposes the universe would be finished. This degenerating situation persisted for about 2,000 million years until a time came when the system had to have the equivalent of a rethink, perhaps an actual rethink.

## But what sort of rethink?

In talking about 'rethink' I am, of course, using anthropomorphic language which would occur to any of us if we had been 'in charge'. We would say 'This cannot be allowed to go on or we have all had it!' I do not justify this language, but I fancy we have no other language to apply to the situation. It was a *fact* that the universe was running down to extinction under the Second Law of Thermodynamics, and entropy was always remorselessly increasing. The universe was *spending its capital*, and the situation was alike to ourselves when each month our bank balance shrinks so we can even calculate when we will 'go bust'.

## Hope from Clerk Maxwell

It was Clerk Maxwell in the nineteenth century who told us that the Second Law of Thermodynamics could be reversed by 'conscious sorting demons'. But what would that mean applied to the running down of the universe? Would it imply that the direction of solar radiation should be reversed and sunlight should be stuffed back into the suns? Or could it mean that 'somehow' there was the possibility of a 'conscious sorting' process of a different kind which would have the same effect? If we consider the most general analogue of the Second Law of Thermodynamics and its possible reversal, it may be as well to recall the example of playing cards given in Chapter 7:

1. Disorder and increase of entropy is like the process of shuffling playing cards, an absent-minded process.

2. The creation of order and reduction of entropy (increase of specificity) is like consciously sorting playing cards in their suits and sequences.

## Entropy and sameness

If we adhere strictly to the concepts of the Second Law of Thermo-dynamics, we find that it is concerned with the elimination of temperature potentials as the following example shows:

If I put ten gallons of hot water into my bath and follow this with adding ten gallons of cold water, then I shall have twenty gallons of tepid water. What I have done is to shuffle hot with cold. Now I can never re-separate that original temperature potential between hot and cold; I cannot unshuffle or re-sort the situation. Why does this matter? Because with temperature differentials between hot and cold I can do useful things: for example, I can use the differ-ential to run an engine, perhaps generate electricity and so forth. But I can do nothing useful with heat at a single temperature.

The central point of the above is that energy is useless when it has lost its *variety*, its temperature differences. I want to concen-trate on this point about variety, for one can redefine the Second Law of Thermodynamics:

The Second Law of Thermodynamics tells us that utility depends upon differences, since utility depends upon an energy flow between differences. Thus entropy (uselessness) is infinite when differences (variety) are eliminated.

When we talk about the universe running down, we do not mean that it will turn to a block of ice but rather that it would have no temperature differences and so no thermal currents could flow. There would be no 'voltage'.

## Specificity and variety

Elsewhere in this book (Chapter 7) we have made the case that the opposite to Entropy is Specificity and that they are even mathema-tically related so that:

Entropy = 1/Specificity, or
Specificity = 1/Entropy.

In the present chapter I have just made the case that Entropy is related to 'sameness' and the lack of potentials of temperature or voltage. It is thus a logical step to assume that Specificity will be the opposite and thus related to variety and potential differences. But I would not be content to rely on such inferential logic, and the case must be made more directly that Specificity is directly related to Variety.

### EXAMPLE — A MATHEMATICAL COCKTAIL PARTY

Imagine that I hold a cocktail party for twenty professors of mathematics each of whom knows exactly the same amount of mathematics so none can teach any of the others any mathematics nor can any learn any more mathematics. It is a state of maximum entropy in the field of knowledge and the professors could only talk about the weather.

But let us imagine that there appears in their midst a youth who knows no mathematics. Immediately all the professors can descend on him and instruct him in the mathematical art. This tells us that what matters in energetics is differences or variety. The fact that the professors had great mathematical knowledge counted for nothing until someone appeared who knew no mathematics.

So we can identify entropy with sameness and specificity with variety.

## Organism and variety

An organism considered as a structure or process comprises a set of parts having different functions. In the human being one has the organs of brain, heart, lungs, liver and so forth, and each of these operates quite differently. Even at the level of individual cells we find equivalent functional variety in the nucleus, organelles and cell wall. We can therefore define an organism as follows:

An organism is a structure of varied parts which maintain their variety-integrity over a lifetime.

An organism is thus a system of high specificity and low entropy, and as Erwin Schrödinger has stated:*

> What an organism feeds on is negative entropy. Or, to put it less paradoxically, the essential thing in metabolism is that the organism succeeds in freeing itself from all the entropy it cannot help producing whilst alive.

So an organism is an actual Maxwell 'sorting demon'.

## Stocktaking

We have the situation:

1. The universe is running down under the Second Law of Thermodynamics so that all life and useful energy would be extinguished into a grey sameness.

2. Clerk Maxwell tells us that the Second Law can be reversed by conscious sorting demons.

3. Organisms are sorting demons.

So the leading question arises:

4. Do organisms keep the universe wound up in spite of the Second Law of Thermodynamics?

If this could be established, then it might be a vital pointer to the nature of LOGOS as that influence which creates the specificity of organism to keep the universe wound up.

---

* *What is Life?*, p. 76.

## The variety of species

We have seen that an organism maintains its specific integrity because it is based upon a system of *parts of specific variety*, and the higher the organism the more this is so.

But organic life comprises a vast array of species — millions of them when one includes animals, plants, insects, bacteria and viruses — and all these interact together to some degree and particularly when some are polar pairs. It was the poet William Blake who ruminated on the fact of *tigers* and posed the question:

*Did He who made the lamb make thee?*

It is difficult to imagine a polar pair so distinct from each other and yet so capable of creating mutual vivid experience. The tiger sees the lamb as a perfect good dinner and the lamb sees the tiger as the Devil Incarnate. Thus when the tiger confronts the lamb we have an explosion of experience, and we might well call this *drama*.

So the interactions between the various species create an ongoing drama, sometimes slight and sometimes intense and that according to the degree of the polar variety involved. The situation is somewhat like an electric battery with positive and negative poles: if one connects a wire between the poles, then a current will flow, but this current will be proportional to the polar voltage. So the tiger-lamb pair constitutes a high-voltage situation, whereas the blackbird-robin pair constitute a low-voltage situation. But I imagine there are no zero-voltage relationships, and I have observed in my own garden quite a wary and respectful relationship as between the blackbirds and the robins: there is no doubt but that there is generated between them a degree of conscious experience.

We ask the question: since we have seen that variety-specificity acts the role of Maxwell's sorting demons, does the fact of the variety of species enhance this trend? Surely there can be no doubt of it. So we demarcate two degrees of specificity:

1. A given organism (species) maintains its living integrity due to the specific pattern of its construction.

2. The variety of species extends this principle into the external world of inter-relationships.

## Sexual polarity and genetic variety (individuality)

The fact that masculine-feminine sexuality is the basis of repro-duction of species is further evidence for variety based on polar specificity. This is not merely a simple polarity, but the genetic combination from both parents ensures that the offspring are different from either and so enormously increasing the magnitude of specificity based on the unique nature of individuality. We know this to be true of mankind, but it must be true of all species down to bacteria which are now known to be sexed.

Thus altogether there is a fourfold system of specific variety through the system of organic structure, environmental relation-ships, sexuality and individuality. The mathematics of such speci-ficity, expressed by multiplying together these four factors, is incalculable and must be assigned the simple description of 'infinite'.

## The consequences of the variety of organic life

Let us consider only the single factor which is related to environ-mental experience and which consists of the encounter between organic individuals. Provisionally we can call this 'experience'.

Under the Second Law of Thermodynamics, entropy increases when any two entities are involved in a blind collision, for example as between two gas molecules. When this occurs in a volume of gas, the result of such blind collisions is a sort of shuffling.

But the encounters between organic individuals are *conscious collisions*, or perhaps a better description is conscious encounters. But they are quite analogous to what happens in a gas under blind chance, with the major difference that the system generates experience and consciousness. When the tiger chases the lamb, there is an expansion of the tiger's conscious experience and the lamb is also alerted into a higher state of consciousness corresponding to fear. So we can state an hypothesis:

The interactions in an organic system ('life') act as a generating station for experience and consciousness.

The point we are pursuing is whether life is a system for the creation of Maxwell's sorting demons — and, if so, can such a system wind-up the universe through reduction of entropy? The answer is by no means obvious, but one clings to the significance of the fact of the generation of consciousness, for consciousness is the essential characteristic for a sorting as distinct from a shuffling system.

The stated purpose of this chapter is to discover whether life has a purpose and if so what it could be. The suggestion is that life is a system for the generation of consciousness and that this may be related to offsetting the running-down of the universe under the Second Law. We pursue this theme further.

# 23

## THE LOGOS CYCLE AND THE GREAT LABORATORY

Nature shows great domestic economy by ensuring that nothing is wasted. What is rejected from one creature is used by another. The prime example is the way in which plants need carbon dioxide but reject oxygen, whereas the animals need oxygen and reject carbon dioxide (see Fig. 23.1). A little study shows that all self-sustaining arrangements have to be of this recycling nature, otherwise some vital material would eventually become exhausted. Since we are searching for some system which may be relevant to winding-up things rather than running-down to exhaustion, then this suggests that we should contemplate a recycling system. We could need to work in at least the following factors:

LOGOS
DNA
Creatures
Experience.

So let us arrange these as a recycling loop as shown in Fig. 23.2. We assume that the cycle has its start in the programming influence of LOGOS, which creates DNA, which creates creatures, which have dramatic experience and which goes back to LOGOS so that the cycle can restart. We might call it the *Life Drama Cycle*.

### The evolutionary Life Drama Cycle

Expressed in this fashion we note that this is a very strange cycle since it is one where LOGOS ultimately 'feeds' on the drama of its own creation. Thus it must be evolutionary if it is to continue to produce consciousness. Let us consider the words of

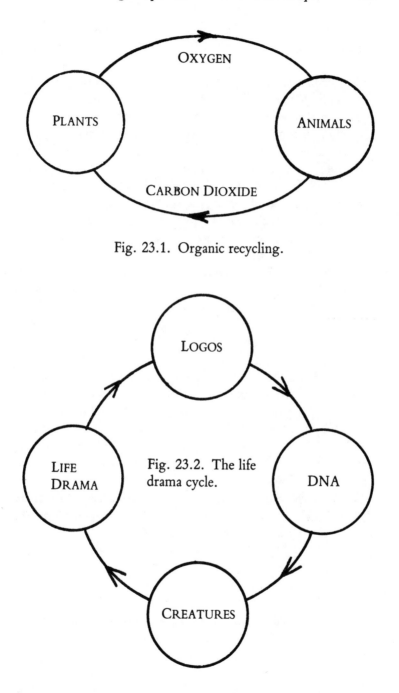

Fig. 23.1. Organic recycling.

Fig. 23.2. The life drama cycle.

Erwin Schrödinger on the general philosophy involved in consciousness:*

> Any succession of events in which we take part with sensations, perceptions and possibly with actions gradually drops out of the domain of consciousness when the same string of events repeats itself in the same way very often. But it is immediately shot up into the conscious region if at such a repetition either the occasion or the environmental conditions met with on its pursuit differ from what they were on all the previous incidences. . . . One might say, metaphorically, that consciousness is the tutor who supervises the education of the living substance, but leaves his pupil alone to deal with all those tasks for which he is already sufficiently trained. . . . Consciousness is associated with the *learning* of the living substance; its *knowing how* is unconscious. . . . Consciousness is a phenomenon in the zone of evolution.

This is a very important and significant statement. We have postulated a Drama Cycle in which life-experience may feed back to LOGOS, but if this is consciousness about life then Schrödinger's interpretation would suggest that only *evolutionary experience* is thus fed back. In this case the Drama Cycle is an evolutionary cycle (or spiral?) in which LOGOS itself is learning. Also note the relationship of this to the general principle of specificity since learning is the *variety* aspect of life, that which stands out as being different from repeating function.

So the purpose of the Drama Cycle is to create and then to observe how life was adapting to its environment through the learning experience of environmental interaction.

---

* *Mind and Matter* (Eng. transl. Cambridge University Press, 1967), p. 102.

## The Great Laboratory

Since LOGOS is responsible for the DNA and creatures, and since it may receive feedback from evolutionary life-progress, then the whole is a progressive laboratory system with organic life itself the laboratory and LOGOS 'the man in the white coat'.

This invalidates Darwin's theory of evolution. To Darwin life was a free-for-all jungle for survival of the fittest, whereas the Great Laboratory concept is of a conscious cosmic purpose and awareness directing progressive experiments. At the present time man is the last in the line of the experiments which started with single cells such as the amoeba, but there is no knowing whether he is the last. Anyone who would wish to know where the Great Laboratory is need only look out of the window at the garden outside.

## A biological fallacy

In molecular biology there is a central dogma that there can be no feedback from proteins to the DNA; it is a one-way command system in which the DNA does all the programming. I agree with this, but some scientists infer that the DNA is independent of environment. This is clearly a fallacy since birds have wings appropriate to an aerial environment whereas fish have fins appropriate to a watery environment. Thus there is a sort of logical feedback of necessity; however, I accept that it is not a chemical feedback, for the chemical feedback must be prohibited to preserve the specificity of a species. The remarkable matter is not the evolution of species by natural selection but the durable specificity of species through aeons. I suggest that the logical feedback is via creaturely experience to LOGOS as the cycle of Fig. 23.2, and then LOGOS may do some re-programming of the DNA.

## *The methods of the Great Laboratory*

The intelligence of LOGOS does not need to be always infinitely exerted. That is demonstrated by the fact that life was originally confined to the relative simplicity of single-cell floating creatures. While it is clear that more complex creatures 'evolved' from the more simple, it could equally mean that such evolution could have taken place in successive steps, quantum jumps. Indeed the mathematical specificity of life (Chapter 14) requires this sort of evolution within the finite known time of organic life.

But now we are beginning to credit LOGOS with being a conscious intelligent entity, and thus it could use the same methods as we human beings do in our constructive efforts. When human beings attempt anything complex, they approach this through a master programme which contains many standard sub-programmes. If one wishes to build a skyscraper, one specifies its materials such as iron girders and sheets of glass from catalogues which imply sub-programming already existing for the production of such materials. It would be very reasonable for such sub-programmes to exist for the Great Laboratory. Thus to specify a dog and a cat could involve the common sub-programmes for

> Four legs and a tail
> One digestive tract complete
> Body hair cover.

After that it could branch into:

| CAT | DOG |
|---|---|
| Tiger eyes to routine A | Normal animal eyes to routine C |
| Shape as sketch B | Shape as sketch D |
| Temperament — independent | Temperament — dependent domestic |

So, in order to create a new species, LOGOS only needs to figure out the *new* evolutionary requirement, call on the card index of sub-programmes for 90 per cent of the structure, and then add the new or modified touches.

No? Why not? Why should LOGOS do it the hard way and think every new species up from scratch? One suspects that it will not be long before we identify sub-programmes in the DNA as to 'four legs and one tail' and so forth.

But how can LOGOS 'observe' the performance of species from which it could decide on evolutionary improvements? In the next chapter let us turn again to Erwin Schrödinger.

# 24

## ERWIN SCHRÖDINGER AND 'THE ONE SELF' (OR MIND)

Erwin Schrödinger, who created Wave Mechanics, wrote several philosophical books, outstanding among which were *What is Life?* and *Mind and Matter.* In these books he introduces the very bizarre idea that there is only one Self (or Mind) in the universe. To some considerable extent Schrödinger's views are based on earlier reflections from Sir Charles Sherrington, the eminent British physiologist, as described in Sherrington's *Man on his Nature.* * Let us start with Sherrington.

### Two eyes but one mind

Sherrington recounts a famous experiment of considerable psychological significance. It is well known that if a moving scene is presented to the eyes but is interrupted by a moving rotating shutter, the scene will flicker until the shutter is making about sixty interruptions a second and then the flickering will cease. This principle is made use of in the showing of cinematograph and television presentations, which consist actually of a series of still pictures. But if the two separate eyes are subjected to a shutter at the rate of thirty a second but alternatively phased, then if the nervous ocular system were joined to a single nerve system one would expect no flicker since there would be a 30 + 30 = 60 flickers a second. But not so. The separate eyes have to be subjected to 60 flickers each, whether in phase or not, for the flicker to vanish.

Sherrington says:

> It is much as though the right- and left-eye images were seen each by one of two observers and the minds of the two observers

* Cambridge University Press, 1940.

138

were combined to a single mind. It is as though the right-eye and the left-eye perceptions are elaborated singly, and then psychically combined to one.

Another proof of the same nature is related to the picture in Fig. 24.1. Sometimes this looks like a set of steps descending from left to right and sometimes like a cornice (or inverted pair of steps) ascending from right to left. If one looks at this, it will sometimes jump from one form to the other (particularly if one winks the eye) but one will *never* see both forms at the same time with two eyes.

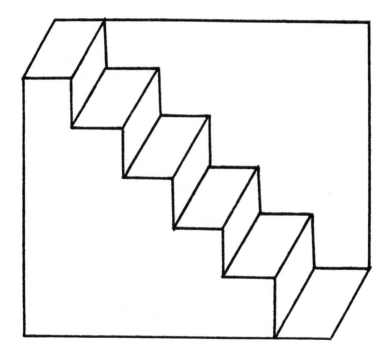

Fig. 24.1. Sometimes these steps appear to present themselves to viewing from above, and sometimes to viewing from below, but never both at once.

Thus the picture is recognised by only one mind, although in two possible forms.

Sherrington uses this argument to indicate that mind and brain (or nervous system) are not the same affair:

> Observers of skill, after devoting patient study to the motor behaviour of single cells, conclude that microscopic single-cell life, without sense organ and without nervous system, can *learn*. There seems no clear lower limit to mind. . . . That the brain derives its mind additively from a cumulative mental property of the individual cells composing it has no support from any facts of its cell structure.

Sherrington thus finds that mind operates in its own sort of unified world and describes this as 'mental space' and 'conceptual space', which is about the same as the idea which we developed in Chapter 20 as to Imaginary Space-Time.

## Schrödinger

Schrödinger writes:

> The reason why our sentient, percipient and thinking ego is met nowhere within our scientific world picture . . . is because it is itself that world picture . . . Here we knock against the arithmetical paradox; there appears to be a great multitude of these conscious egos, the world however is only one. . . . The several domains of 'private' consciousness partly overlap. The region common to all where they all overlap is the construct of the 'real world around us'. There are two ways out of the number paradox. . . . One way out is the multiplication of the world in Leibniz's fearful doctrine of monads; every monad to be a world by itself, no communication between them. . . .
>
> There is obviously only one alternative, namely the unification of minds or consciousnesses. Their multiplicity is only apparent, in truth there is only one mind.

It is an organic concept and is paralleled in the way that each cell of our own body contains the same DNA and 'knowledge of the whole'. Schrödinger quotes an analogy from the thirteenth-century Persian mystic Aziz Nasafi:

> On the death of any living creature the spirit returns to the spiritual world, the body to the bodily world. In this however only the bodies are subject to change. The spiritual world is one single spirit who stands like unto a light behind the bodily world and who, when any single creature comes into being, shines through it as through a window. According to the kind and size of the window, less or more light enters the world. The light itself however remains unchanged.

At this stage Schrödinger goes on to describe the ocular findings of Sherrington as outlined at the beginning of this chapter. He also cites the general fact that we never find consciousness in the plural as an actual experience. It is always singular, and even in the case of the schizophrenic the changes of consciousness and personality are sequential and replace each other in time:

> The overall number of minds is just one. I venture to call it indestructible since it has a peculiar time-table, namely mind is always *now*. There is really no before and after for mind. There is only a now that includes memories and expectations. But I grant that our language is not adequate to express this, and I also grant, should anyone wish to state it, that I am now talking religion, not science — a religion, however, not opposed to science, but supported by what disinterested scientific research has brought to the fore.

So Schrödinger brings us to the frontiers of mystery, as did Sherrington, with the mystery of mind as distinct from brain, and with its strange capacity for integrating experience *now*.

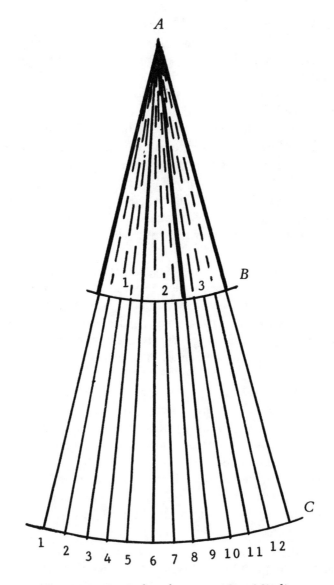

Fig. 24.2. Optical analogue to 'One Mind'.

## The optical analogue of Schrödinger's 'One Mind' proposal

One can make an optical analogue of Schrödinger's One Mind proposal (calling on the statement of that Persian mystic), as shown in Fig. 24.2. This only deals with how it might operate through two organic levels, but in principle it can operate through an indefinite number of levels. The analogue is:

1. There is a central source of Mind or Intelligence which contains all knowledge and knowhow, at $A$.

2. The central source radiates this omniscient knowledge in a spherically expanding wave into Actual Panoramic Space-Time.

3. It encounters a first organic world plane at $B$, which filters from it that knowledge appropriate to its level. Three 'beings' are shown at this $B$ plane as to $B_1$ $B_2$ and $B_3$. Note that the $B$ plane does not use the finer structure of knowledge (shown with broken lines or dashes) but only that indicated by the solid lines.

4. The fine structure passes through the $B$ plane to the C plane where it now relates to beings $C_1$ to $C_{12}$, and provides them with knowledge appropriate to that plane.

If by analogy we apply this to man, then we can imagine that $A$ corresponds to a person as a whole, the $B$ plane corresponds to organs, and the $C$ plane corresponds to cells. But on a different scale we might imagine that $A$ corresponds to the Biosphere (organic life as a whole), the $B$ plane to a species (such as man), and the $C$ plane to individual human beings.

The analogue is certainly of the correct 'shape' to accommodate Schrödinger's notion, with an omniscient light shining intelligence from $A$ through a succession of planes but with progressive specialisation (fine structure) of the knowledge relevant at a given plane.

## The feedback of organic experience

This construction shows how the differentiated experience of organic life could feedback to LOGOS at the origin and how every single individual could be interrogated for its experience.

# 25

## MAN AND LOGOS

### 'Logos' and the mental aspect of DNA

We have defined LOGOS as 'that which is behind the DNA'. But one aspect of the DNA must relate to the central species characteristics, and in the case of the human being, that we hope is what, that we mean by 'Homo Sapiens'. So one aspect of human DNA must relate to the psychic aspects of man. There is some tradition for this in the concept of the Conscience (Old English: Inwit). So the schematic of man could be as shown in Fig. 25.1 in that there is an aspect of the DNA which is programming for his body and another aspect of the DNA which is programming for his psyche: and both related to a ghostly ultimate programming by LOGOS. This analysis agrees with Schrödinger's view about there being only one mind because it suggests that behind each of us, and coming via the psychic aspect of the DNA, is this mighty intelligence of LOGOS.

### The problem of language

We already know from molecular biology that the DNA has a language of its own, and we can hardly expect this to be the Queen's English because it must relate to all nations and races. This cosmic language which appears to have been unchanged for the whole stretch of organic life must be some sort of cosmic language which is totally coded in relationship to Earthly languages.

So if LOGOS were to try to tell us something, how can we understand the message?

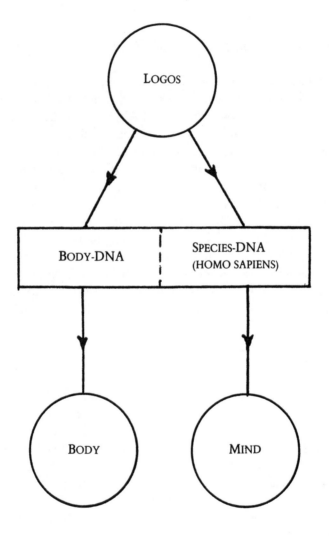

Fig. 25.1. The two aspects of the DNA.

## *A hint from mystical physics*

All the major developments in physics of this century have been of mystical origin, with the outcome generally being some new mathematical formulae. As Sir James Jeans has stated (Chapter 3):

> Nature seems very conversant with the rules of pure mathematics. . . . The universe appears to have been designed by a pure mathematician.

So does LOGOS use some esoteric symbolism in conveying messages to human beings via the psychic aspect of the DNA? If one considers the mystical experiences of the great pioneering scientists, then in each case one can deduce that they experienced mystical revelation. One might quote:

Planck and quantum theory
Einstein and relativity theory
De Broglie and matter-wave equivalence
Schrödinger and wave mechanics
Heisenberg and the uncertainty principle
Pauli and the exclusion principle.

Not one of these was reasonable or common-sense, but all were true and they *worked*. They were the result of 'inspiration'.

## *Inspiration in science*

A writer who extensively studied this subject was Arthur Koestler, and extracts in this section are from his book *The Act of Creation*:*

> *Henri Poincaré*, who is credited with having forecast relativity mathematics and theory:

---

* Extracts reprinted by permission of A. D. Peters & Co. Ltd.

'Just at this time I left Caen, where I was then living, to go on a geologic excursion under the auspices of the school of mines. The change of travel made me forget my mathematical work . . . we entered an omnibus to go to some place or other. At the moment when I put my foot on the step the idea came to me, without anything in my former thoughts seeming to have paved the way for it, that the transformations I had used to define the Fuchsian functions were identical with those of non-Euclidean geometry. I did not verify the idea; I should not have had time, as, upon taking my seat in the omnibus, I went on with a conversation already commenced, but I felt a perfect certainty. On my return to Caen, for conscience's sake, I verified the result at my leisure.'

*Karl Gauss*, the pioneer of magnetism:

'At last two days ago I succeeded, not by dint of painful effort but so to speak by the grace of God. As a sudden flash of light, the enigma was solved. . . . For my part I am unable to name the nature of the thread which connected what I previously knew with that which made my success possible.'

*August von Kekulé*, who discovered the ring of six carbon atoms which is the central structure of organic chemistry:

'I turned my chair to the fire and dozed. Again the atoms were gambolling before my eyes . . . long rows sometimes more closely fitting together; all twining and twisting in snakelike motion. But look! What was that? One of the snakes had seized its own tail and the form whirled mockingly before my eyes. As if by a flash of lightning I awoke. . . . Let us learn to dream, gentlemen.'

There are many more examples of sudden scientific inspiration when a person is in a relaxed state or even day-dreaming.

## *What is inspiration?*

Koestler took the view that inspiration can occur only when a problem has been worked on from many points of view so that it is 'ripe' for a solution, and my personal experience agrees with this. One does not achieve illumination without having first tilled the ground. But it appears to be the common experience that 'flashes' are initially in the nature of a non-linguistic and non-image *emotion* of certainty, which only later shows a pattern of structure and detail. The nature of the emotion is simply an intense feeling that a problem has been solved, but the sense of the origin of the solution is non-personal and transcendental. It is as though the sense of wonder extinguishes 'myself'.

## *Is emotion 'compressed data'?*

It is common to consider emotion as though it were a category distinct from thought or knowledge, a fundamentally different faculty. But, as the following reasoning suggests, this may not be so.

We are used to consider that a thought is lucid and digestible provided it endures for a given length of time, the sort of time-duration we associate with 'now' — perhaps a few seconds. But we also require that such a lucid thought is *a single thought* such as 'two and two makes four', and we cannot cope with it if it is too complex.

But if we refer to the end of Part 1, the ultimate conclusion was that 'reality is organised mind-stuff'. The schematic of this is Fig. 5.1 where ten inter-relationships, each of which could be a simple idea, are shown as a single organic whole. Thus it is suggestive that an emotion is simply such an organic idea which has several contributory facets or points of view, all of which are valid. Indeed, one would expect that important ideas are organic. So an emotion is the collective signature-tune of a complex idea and, although the emotion may be stating 'problem solved!', it then needs analytical time to unscramble and analyse the parts.

The problem of an emotional understanding is that it is complexified data. Let us illustrate this by analogy to an accomplished man-made technology of compressed data.

Consider a tape-recorder on which has been recorded the poem *Daffodils* (see Chapter 10). If one plays that tape recorder, the sequence of words makes lucid sense:

*I wandered lonely as a cloud* etc.

But now speed that tape up by a factor of ten times and, although the meaning is still there it no longer makes sense, everything has run into everything else. I can no longer decode the specific meaning of

*'I'*
*'Wandered'*
*'Lonely'*
*'A Cloud'*

Thus the organic whole of the statement is lost. This means that we human beings can only understand statements which are recounted in a time-scale which is proper for our understanding. But although I have given an example of data compression (which is now used in international communications to speed up messages), there is the further likelihood that organic ideas (several points of view) are not sequential but are simultaneous (see Fig. 5.1) and totally overlap each other. So ideas which come to us as emotional flashes cannot be immediately decoded,even though there is a single overall message.

Let us attempt to put all this together:

The great breakthroughs in science were almost all of a mystical nature in which an emotion of 'problem solved' preceded the solution in Queen's English or mathematical symbols. This suggests that higher levels of thought are organic complexes and that at first we only experience these as an emotional flash of certainty.

## Navigating emotional organic thought (from 'Logos'?)

After many years of experience of this sort of mental phenomena I have learned to recognise certain flashes of positive emotion as the forerunners of 'problem solved', to the degree that one can hold the taste of such emotions until they reveal their contents. It is suggestible, in view of the suggestibility that LOGOS operates through the mental aspect of human DNA, that at such moments we are unified with LOGOS in a way that is compatible with Schrödinger's 'one mind' hypothesis.

# 26

## A NEW THEORY OF EVOLUTION

Organic life started in the oceans, which are still by far the greatest reservoir of life. This initiation is to be expected, since the formation of the original single-cell plants requires a fluid medium for the mobility of transportation of the various chemicals needed. All the chemicals needed for life are dissolved in sea-water.

The history of life proceeds from the simple to the complex, and the question is: what is the main category to which such simplicity or complexity relates? If one takes the two extreme ends of the evolutionary spectrum, from amoeba to man, the main difference appears to be the size of the two corresponding significant environments. The amoeba lives in a 'world' to be measured in 'millimetres', while man lives in a world to be measured in light-years of distance. Man does not experience personal mobility to the degree of light-years, but his sense organs and mind create for him a huge 'world at a distance'. This category could be called environmental intelligence, the world with which a creature is in communication.

## *Relative environmental intelligence as the criterion for evolution*

Such a simple analysis enables us to define a new criterion for organic evolution:

Evolution is that process which increases the effective world size within which a creature can act or communicate or understand.

Evolution will thus be partly a matter for the proliferation of the receptive senses and also of expansion of the associated inner mind. Different senses enable creatures to communicate with different worlds such as:

Eyes.   The visual world
Ears.   The world of sound
Nose.   The world of smell
Mouth.  The world of taste
Touch.  The world of direct physical contact.


## Man and the Universe

It has been reckoned by astronomers that man, aided by his telescopes, can already survey over half the total universe; the invisible residue vanishes into far outer space with the velocity of light, and the approach to this condition is marked by the red-shift. So for man there are no major new physical worlds to discover; his problem is to make sense of what he has already discovered.

Of the five human senses mentioned above, by far the most important is that of vision, for it is this which has opened up man's world survey. But, as just indicated, the problem of man is not to see more but to *understand* more. Thus we can come to a more refined definition of evolution:

Evolution is that process whereby an organic creature can understand more about its environment, whether in a quantitative or qualitative sense, i.e. whether in terms of size or of specificity.

In this fashion evolution is the *cultivation of intelligence*, and perhaps particularly through the intensification of visual acuity.


## The role of sunlight for evolution

The key condition for the evolution of organic intelligence is sunlight. Sunlight is a radiation whose properties extend from that of mere energy to that of conveying highly sophisticated information. The purely energetical aspect we note in the process of photosynthesis of green stuff; this is the most elementary starting-point for life. But sunlight as 'light' has remarkable properties in that it is *information* modulated (by superimposition) by any forms on which it falls. Let us imagine an elephant standing in sunlight. The approaching rays know nothing about the elephant, but as soon as they strike the elephant and are reflected from it, then the reflected rays contain information about the form of the elephant. This is proved by the fact that the emergent reflected rays can be intercepted by a lens (of eye or camera) and we have an image of the form of the elephant. This image is inherent in the rays of light coming from the elephant; all the lens does is pick out a particular cone of rays to form the image. Note that the image can be equally secured by a pin-hole, and a pin-hole itself can hardly be responsible for the form of the elephant.

In this fashion reflected sunlight produces an almost infinite number of images and thus creates an information system, a system of potential knowledge; the role of organic senses is simply to conduct this information to particular individual creatures 'receiving sets'. But the information broadcasting is due to sunlight modulated by terrestial shapes. So the illuminated biosphere is an information complex for its creatures.

## The progressive evolution of creatures in the solar information complex

Given this information system, having its origin in sunlight and throwing up the form of the world, one can then trace an organic ladder of relative creaturely intelligence as:

1. Single-celled creatures in the sea have their intelligence potential limited by the extent of their environment, which is 'millimetres'.

2. When these develop into fish, then their intelligence is limited to their field of mobility and vision. In water this is limited to 'metres'.

3. When marine animals crawl out on to the littoral or beaches, then the clear air compared with the dark ocean water immensely

expands their potential territory for communication. But since such creatures tend to be flat-landers, we might think of their world size as $X^2$ where $X$ is the range of vision of mobility and $X^2$ is the corresponding territorial area. Typically $X$ may be 'kilometres'.

4. The evolution of animals into birds adds a third dimension of height, and so their territory is to be represented by $X^3$ where $X$ may average several kilometres and in the case of migratory birds their world volume can be almost of planetary proportions.

5. Finally we come to man who, armed with his telescopes and microscopes, also has a world size of $X^3$ where $X$ may be light years. But because of the properties of human imagination and memory, the world size is $X^3T$ where $T$ is the lifetime of understood experience.

The above is intended to be a very crude representation of stages of evolution on a scale of relative intelligence, with a major quantum jump as life crawls out from the obscuring surrounds of salt-water to the clarity of the atmosphere and with acuity of vision the most important channel for data inputs.

## The development of human controlled imagination

The 'size' of the human mental world is enormously increased by the power of human memory and imagination. In the first place this enables separate percepts to be integrated into common concepts and possibly ultimately into a single world model. But of equal significance is the ability of the imagination to make reasonable guesses as to the nature of reality and, armed with such hypotheses, to go looking for experimental proofs. The ultimate development is a world model which is not even essentially referable to the world of physical vision, since a world of form can be transformed into a world of symbols and their relationships, as in mathematics or logic.

## An overall philosophy of evolution

We have sketched in a new concept of organic evolution which is a ladder of intelligence, the intelligence relating to thought, communication and behaviour with reference to the environment. It is the process of enlarging the size of organic *home* — that aspect of the universe with which we feel familiar, which we can understand and to some extent manipulate.

If there is a basic philosophy, it is a sort of 'crawling towards the light' with the Sun the prime attractive source.

## Where Darwin under-estimated the situation

It would not be fair to Darwin to make out that 'he had it wrong'. For Darwin was without the modern scientific knowledge of molecular biology and the astonishing specificity being revealed in that subject. The two significant modern developments are:

1. The rigid nature of specificity such as that of haemoglobin (Chapter 14) and represented by the 'infinite' number of $10^{650}$.

2. The realisation that chance mutations will essentially, and in almost all cases, be unfavourable for life. As Erwin Schrödinger has stated:

Frequent mutations are detrimental to evolution. . . .

Chance mutations, which are indeed possible, are an interference with the natural order of things and on balance are equivalent to a Death-Wish. We have seen that life is dependent upon specificity, and so we must not expect that chance can do anything more than increase disorder in the situation. This does not rule out favourable mutations, but chance does not suggest their favourable overall balance.

Where Darwin went wrong was in underestimating the rigid degree of specificity to 'magical' magnitudes. Darwin's theory was common sense, but common sense does not apply in dealing with situations where the specificity is $10^{650}$ or greater. Darwin's theory was excellent logic in a universe bereft of either DNA or LOGOS.

# 27

## IS DNA PROGRAMMED FROM THE SUN?

In this book we have noted the emerging significance of the Sun in encouraging organic life on Earth. The principle points have been:

1. Sunlight provides the prime energy for sustaining life through the photosynthesis in green vegetation (Chapter 12).

2. The Sun provides the conditions favourable for organic life through several means whose specificity is to be measured by the number 700, the odds-on specificity (Chapter 21).

3. The direction of the evolution of life appears to be a 'crawling towards expansive intelligence', with sunlight giving the world meaningful form (Chapter 26).

Thus it would not be a surprise if the last remaining enigma, the source of the DNA, should not also be a solar property. We have associated the source of the DNA with LOGOS, and there is a long religious tradition which associates deity with the Sun — the Sun God.

### Is the Sun intelligent? Catalytic processes

The three conditions just given suggest that 'the Sun knows what it is doing' with reference to organic life. But can we find less inferential and more direct proof of solar intelligence?

There is one possibility, namely the fact from chemistry that the most economic and efficient processes are *catalytic*, indeed to the degree that they can be considered as Maxwell's sorting demons and that the major process in the Sun is catalytic.

It is now known from organic chemistry that, after the DNA, the most intelligent reactions are connected with enzymes which are proteins that enter into chemical catalytic reactions but come out again unscathed, i.e. without increase of entropy. Jacques Monod, a leading molecular biologist and Nobel Prize winner, wrote in his book *Chance and Necessity*:

> Each of the thousands of the chemical reactions that contribute to the development and performance of an organism is provoked electively by a particular enzyme protein. . . . In the organism each enzyme exerts its catalytic activity at only one point in the metabolism. . . . These phenomena, prodigious in their complexity and their efficiency in carrying out a present programme, clearly invite the hypothesis that they are guided by the exercise of somehow 'cognitive' functions.

What the enzymes do is, as it were, to hold two chemicals with each hand and get the two chemicals themselves to hold hands together; the enzyme, having done its introductory work, then withdraws from the scene. The situation is not unlike that of a good hostess holding a party and introducing people to each other and then moving on when two people become locked in conversation. This type of process is called catalytic.

Catalysis is a highly intelligent process, although I shall not attempt to define 'intelligent' except to say that it behaves as though it were conscious (as does an electronic computer).

## The carbon-cycle catalytic process in the Sun

In the Sun too there is a major catalytic process, and that is the way by which energy is generated by the conversion of hydrogen to helium via a catalytic process in which carbon, nitrogen and oxygen are the catalysts. This catalytic cycle, discovered by H. A. Bethe and elaborated on by W. A. Fowler, is:

$C^{12}$ plus $H^1$ converts to $N^{13}$
$N^{13}$ plus $H^1$ converts to $C^{13}$
$C^{13}$ converts to $N^{14}$
$N^{14}$ plus $H^1$ converts to $O^{15}$

$O^{15}$ plus $H^1$ converts to $N^{15}$
$N^{15}$ releases $He^4$ to convert to $C^{12}$.
(This is where we came in!)

Thus hydrogen converts to helium through a cycle involving carbon, nitrogen and oxygen, and all through the cycle various amounts of energetical radiation are being given out. So in this cycle carbon, nitrogen and oxygen act like Maxwell's sorting demons, taking part in organising catalytic activity but emerging unscathed.

So if catalytic reactions are 'intelligent', then the Sun would appear to be intelligent.

## Solar radiation and information transmission

So far we have given three arguments why the Sun *ought* to be the source of the DNA for the sake of logical elegance, and we have added a single separate fact that the Sun manifests intelligence inasmuch as the presence of catalytic processes indicates the intelligence of Maxwell's sorting demons. But now let us turn to the nature of solar radiations and ask ourselves the question:

Could solar radiations have the information-handling capacity corresponding to the astronomical data content of DNA?

It has been estimated that the information content of the DNA is about equal to 20,000 long books, say of 500 pages each, and thus the totality is about 10,000,000 'pages'. On each page of a book there are about 2,000 letters, so the totality equivalent of DNA is thus $2 \times 10^{10}$ letters. Let us assume that the sequential code is in binary form of (say) 16 different digits for each letter, so that the information totality is now about $3 \times 10^{11}$ binary digits. To transmit a binary digit with clarity over a radiation system such as radio or light requires about 30 waves to differentiate as between the bit 1, the bit 0 and a 'space', so that we need a radiation system of some $10^{13}$ waves.

Some of the main solar radiations have the following frequencies in cycles per second (on the right side are given the times which would be needed to transmit the total DNA message involving $10^{13}$ waves):

| Radiation | Radiation frequency | DNA message time |
| --- | --- | --- |
| Infra-red | $10^{13}$ | 1 second |
| Visible light | $5 \times 10^{14}$ | 1/50th second |
| Soft X-rays | $10^{16}$ | 1/1000th second |
| Hard X-rays | $10^{19}$ | 1 microsecond |
| Gamma rays | $10^{20}$ | 1/10th microsecond |
| Cosmic rays | $10^{22}$ | 1/1000th microsecond |

Thus, even using the slowest radiation of infra-red, the DNA message would only take one second to transmit, and it would then take 8 minutes to traverse the 92 million miles from Sun to Earth. It is feasible even if it is not plausible. If I had to guess which was the radiation carrying DNA information to Earth, it would be gamma rays since these are the main energetical product of the carbon cycle referred to earlier. But it would be a very wild guess. On balance I would rely on the earlier logical argument that if the Sun provides three main support aspects for organic life on Earth, it would be very odd if it did not provide the fourth, the programming of the DNA. In that case LOGOS would be something like 'the Sun God'.

# 28

## SPACE, LOGOS AND CREATION

In the last chapter we considered the relationship of organic life to the Sun and whether the DNA could be programmed from a LOGOS associated with the Sun. We now consider the creation of the whole universe, i.e. the totality of Suns and matter-energy.

### *The strange properties of space — curved space*

The idea that space is just 'nothing' (the Void) was corrected by the development in the last century of non-Euclidean geometry. This played a vital part in the development of Einstein's special and general Relativity. Although there is 'nothing' in space, space none the less has very important properties. One of these properties is that mass in space 'curves' the surrounding space. It is general Relativity which tells us that planetary orbits are not due to gravitational attraction between the Sun and planets 'at a distance' such as the Earth but that a planetary orbit is simply the curvature of space round the Sun. So the Earth might well consider that it is moving in a straight line but actually it is moving on curved 'railway lines' created by the curvature of space round the Sun. The main proof was that the same properties should cause light to be bent as it passes close to a large mass such as the Sun. This has been proved by the bending of starlight by the Sun, as can be observed during an eclipse. So we can summarise:

MASS CURVES SURROUNDING SPACE.

### *Einstein's General Relativity*

Einstein's General Relativity is a very remarkable theory because it can interpret mass-energy as to an equivalent distortion of space-time. This is known as Einstein's matter-tensor, and I quote from Condon and Odishaw's *Handbook of Physics* (McGraw-Hill, 1958, page [1] 122):

> Einstein equated the tensor $T_{ik}$ with the matter tensor of theoretical physics and thus obtained a purely geometrical interpretation of mass, energy and momentum. Mass density or energy can thus be conceived in terms of the curvature of a four-dimensional Riemannian manifold. The Riemannian curvature is particularly high at such portions of space-time where there is matter, while in empty space the tensor $T_{ik}$ vanishes.

All this means is that if mass curves space, then mass can equally be defined mathematically by the curvature of space. This could well be called the Law of Mass-Space Equivalence.

## The hypothesis of creation

Einstein thus made clear that mass ('matter') was the equivalent of curved space, but this suggests two alternative interpretations:

1. If one assumes mass, then space is curved around it. Or

2. Given curved space, then one will find mass at the focus.

Which is the basic truth? Or is it a chicken-and-egg situation? What is interesting is the following almost certainly true hypothesis:

If, by some unknown agency, space were to be curved, then this would create matter at the focus of the curvature.

## The curvature of space by thought

Let us remind ourselves of those words of Sir James Jeans, the astronomer, quoted in Chapter 3:

If the universe is a universe of thought, then its creation must have been an act of thought.

The idea that thought is form in space is very ancient. It is explicit in Plato and was developed by Kant and more recently by Hinton and Ouspensky. The logic is very simple:

1. Thought is form.

2. Space, whether real or imaginery, is the medium for form.

3. Therefore, thought is the shaping of space.

This logic seems to be comprehensive whether it is applied to external objects, internal images, the letters and words of an alphabet, or mathematical or other symbols. Shaped forms are the currency of that drama we know as life and experience. But if the universe could be created by the shaping of space, is there an agency capable of such an activity? For this we refer to LOGOS.

## The space-shaping power of 'Logos' inferred from molecular biology

If LOGOS had the power to shape space, then that could account for creation. In Chapter 14 it was shown that the specificity of the haemoglobin molecule was to be represented by the number $10^{650}$. But the haemoglobin molecule is a *shaping* of organic elements. Haemoglobin is based upon molecular shaping, first in DNA and then into proteins such as haemoglobin. This is *super-shaping*, a work of art. The causal agency for this we have named LOGOS because the logic of the situation required such a programming agency. But if LOGOS could shape affairs in space in this fashion at the molecular level, then presumably it might also do so at the macro-level within astronomical universal space.

For reasons which will be given later in this chapter it will be suggested that physical creation is a nul-energy event, and thus the problem becomes simplified as follows: would LOGOS need a greater scope of *intelligence* to create the whole universe compared with the scope of the specificity of haemoglobin, which in Chapter 14 we noted was represented by a number in excess of $10^{650}$? But to account for the matter-energy of the universe only involves about $10^{80}$ hydrogen atoms, and even allowing for an unduly high specificity of a hydrogen atom at, say, $10^{5}$ (to specify the parameters for the structure of hydrogen), one is still at a total low relative level of specificity: $10^{85}$.

It is a very strange thought that the whole universe is a relatively trivial construction in terms of specificity (i.e. order or anti-entropy) compared with a single molecule of haemoglobin. It suggests that we should not be too awed by the immensity of the universe with its light-years and parsecs, but that we should rather concentrate our wonder on the facts of organic life on Earth (or elsewhere).

## Creation as a nul-energy event by catalysis

But we have neglected one factor — the total mass-energy equivalence of the universe, which is all but infinite. However, one thing which we recall from our consideration of the Second Law of Thermodynamics is that 'energy' is virtually meaningless unless we take into account temperature differentials or specificity (anti-entropy). So 'infinite energy' is useless if it is all at the same temperature — under which conditions it can almost be disregarded. However, there are thermal temperature differences in the universe, and so 'energy' must be accounted for.

Even so, I make the case that creation is an event which does not call for any sort of original energy since the process of creation is *catalytic*; and that a causal LOGOS enters into creativity but re-emerges untouched (see Chapter 27 on the intelligence of catalytic processes). The hypothesis is that creation takes place in the following steps:

1. LOGOS manifests creative thought which curves space to create matter such as those $10^{80}$ hydrogen atoms which then partly condense into stars.

2. The created matter has around it the corresponding curved-space field produced by its mass (General Relativity theory).

3. The curved-space then maintains the focal mass which, as it were, is a sort of LOGOS-memory.

4. LOGOS withdraws from the situation, having acted only as an initial creative catalyst.

## The lock-on situation

In such a situation no primary causal energy is required, and the created mass locks on (by feedback) to its formative space-curvature (see Fig. 28.1). The mass-energy situation is then a self-contained polarity as between mass and space curvature, each mutually maintaining the other and thus acting as LOGOS memory. We have the parallel in our human storage of memories which involve no energy to maintain themselves. Our memories are 'set up' by an original experience of intelligence and are then stored for the rest of our lifetimes, surfacing sometimes 'now' and sometimes in dreams but not calling for energy for their maintenance. In human life it is 'experience now' which is the creative catalyst, and it has its exact correspondence in the way that LOGOS can originate creation in an instant and then withdraw as a catalyst unscathed.

## Nul-energy universes

The idea that a universe can be created without energy expenditure is not new and is basic to the thoughts of science about matter and anti-matter. Also note the logical impossibility of some sort of 'original energy' being required to provide the energy of the universe — for where would that 'original energy' come from? One would be led into a series of regressions. The creative act involves not energy but *specificity*, i.e. Maxwell's sorting demon — LOGOS.

An analogue of the principle is shown in Fig. 28.2. On the left

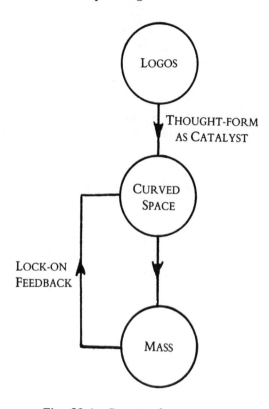

Fig. 28.1. Creation by space curvature.

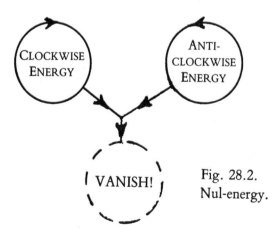

Fig. 28.2.
Nul-energy.

is energy spinning clockwise and on the right is energy spinning anti-clockwise. Bring two such contra-rotating energies together and they both vanish! Similarly take 'nothing' and cause a complementary polarity and one has energy and in the case under consideration the complementarity is that between curved space and focal mass-energy under general Relativity laws. The situation of complementarity has been observed in the laboratory as applying to the electron, the proton and the neutrino and their polar opposites.

Reality in the physical world consists of the establishment of complementary opposites, and I suggest that the basic system is that as between mass and curved or wrinkled space. Support for this point of view seems to come from the duality as between light as a waveform and light as a particle, the photon. The waveform certainly appears to be a shaping of space-time while the photon is the complementary 'substance'.

# 29

## CONCLUSION: THE STUFF OF THE WORLD IS MIND-STUFF

This book began with a statement from Sir Arthur Eddington that 'the stuff of the world is mind-stuff'. Eddington went on to say:

> The idea of a universal mind or Logos would be, I think, a fairly plausible inference from the present state of scientific theory.

Supporting him was Sir James Jeans with:

> If the universe is a universe of thought, then its creation must have been an act of thought.

Their main reasons for coming to such conclusions was the fact that, when physics is explored to its depths, one comes across a world of mathematics rather than a world of 'things'. Similarly, in modern biology we have seen (Part 3) that beyond biochemistry we come to a world of information and literary logic in the DNA, and if one wishes to enquire 'what is behind the DNA?', there is little choice but to propose a similar Logos. It would seem that the developments in biology are even more suggestive than those from physics in confirming that 'the stuff of the world is mind-stuff', for while only some of us can grasp a basic mathematical reality, we can all understand the nature of language. So it would appear that life in the universe, and perhaps the universe itself, is a verbal statement, and that St John had it right:

IN THE BEGINNING WAS THE WORD.

## The destruction of Darwinism

I am not the first to have attacked Darwinism. Others such as Sir Fred Hoyle and Professor Wickramasinghe* have pointed out the statistical impossibility of Darwinism, even if the mechanism of natural selection were to be feasible in principle. But possibly my own proof (Chapter 14), based on calculations of the specificity (improbability) of the protein haemoglobin at $10^{650}$, is novel since it is based on numbers of an exact nature. Anyone with a pocket electronic calculator and access to a modern book on biochemistry can prove for his or her own satisfaction that haemoglobin has the utterly improbable specificity of $10^{650}$, and common sense tells us that Darwinism could not evolve such specificities in a universe which is only $10^{18}$ seconds old. When we consider the specificity of T4 phage DNA at $10^{78,000}$, then the conclusion is final: DARWINISM WAS WRONG. I have not stated the total case against Darwinism, but if the reader is interested I can recommend the book by Hoyle and Wickramasinghe, which quotes specificities in nature of $10^{40,000}$.

But although Darwin considered that organic species evolve by natural selection, he had to retain the concept of God for the creation of original prototypes, and thus he kept within the bounds of 'life comes from life comes from God'. Some scientists have tried to get around this by advocating that life can be spontaneous based upon ultra-violet radiation impinging upon a primeval soup, and there is laboratory proof that some amino-acids can be so created. But amino-acids themselves are not life which requires the combination of amino-acids in specific patterns and sequences. So while laboratory soups can produce what may be considered to be cosmic 'letters', they cannot produce cosmic language of the nature of DNA. This is not a trivial difference since the significance of 'letters' only requires that they be different from each other, unique marks, and anybody can concoct his own alphabet. Humanity has alphabets galore, but meaning only begins when the signs of an alphabet are put together in specific

---

* Fred Hoyle and N. C. Wickramasinghe, *Evolution from Space* (Dent, 1981).

combinations which we recognise as words and sentences.

The significance of Darwin being wrong is immense, for it implies the admission of the *supernatural* into science. There is no known procedure whereby specificities such as the $10^{650}$ of haemoglobin can be explained without introducing a supernatural intelligence into science, an intelligence which can ignore statistics and create unique purpose. Furthermore, this same situation dethrones the Second Law of Thermodynamics from its pre-eminent position in physics, and permits the credibility that this Law can be reversed by the quiet voice of Clerk Maxwell and his 'conscious sorting demons'. Put simply, GOD EXISTS.

## Hoyle* and the Panspermia theory

If life is not due to evolution from a primeval soup, then how does it originate? It would seem that the modern major alternative explanation is the Panspermia theory. This theory was introduced in modern times by the Nobel Laureate S. A. Arrhenius, although it is a very old idea and dates back to the Greek philosopher Anaxagoras in about 450 BC The *Oxford English Dictionary* defines 'Panspermia' as 'the biogenetic theory that the atmosphere is full of minute germs which develop on finding a favourable environment'. The word means 'seeds everywhere' or 'universal seeds'.

As I understand him, Hoyle looks with some favour on the Panspermia idea but with the analysis:

1. Life comes from outer space.

2. Life from outer space is in the form of DNA seeds.

3. The DNA seeds fly through space at very high speeds due to radiation pressure from Sun and stars.

4. The DNA seeds may be subjected to hostile attack in space

---

* The term 'Hoyle' should be interpreted as relating to the joint work and publications of Sir Fred Hoyle and Professor N. C. Wickramasinghe, in their book *Evolution from Space* and previous books *Lifecloud* and *Diseases From Space*.

from ultra-violet (and other) radiations, but this may be mitigated by the carbonisation of the outer surfaces of the DNA to provide a radiation-protective coating of graphite.

But while this theory is of considerable interest in suggesting that seeds from outer space replace any question of spontaneous life generation in terrestial primeval soup, it does not deal with the still main problem of how DNA seeds are created. It simply shifts the problem from Earth to outer space.

At this point Hoyle introduces the principle of *cosmic intelligence* for his First Cause and even suggests that this may use an intermediate technology based on silicon in a similar fashion to that whereby we use silicon chips in computer technology. But ultimately he has to find a 'programmer', and he finds this in a hierarchy of intelligence which has its analogue in a hierarchy such as God-archangels-angels. It is a big jump to make and the last chapter of their quoted book is called 'Convergence to God'.

So Hoyle at this point has to introduce God. In this he is in good company for in my extensive reading about philosophical scientists I have found few, from Einstein to Schrödinger, who at some stage or another did not have to introduce God. The critical moment is when one finds proofs of an intelligence which exceeds human intelligence, and in this book the critical point was the realisation of biological (improbable) specificity. Presumably the most elegant scientist is the one who can go farthest down the scientific road until at last he has to declare: 'I give up. GOD EXISTS.'

At the same time one must admire that minority of major scientists to whom this was anathema. Jacques Monod, molecular biologist and Nobel Laureate, wrote:

Man at last knows that he is alone in the unfeeling immensity of the universe, out of which he emerged only by chance.

His dying words to his brother were: *'Je cherche à comprendre'*, 'I am trying to understand.' One must admire such men who have the nerve to believe in science but not that GOD EXISTS.

## The author's interpretation — the shaping of the void by cosmic intelligence

Perhaps the weakness of the Panspermia theory is that it is a little complicated in three ways. First, it depends on a universal scattering of DNA seeds, and relies on luck to find a germinating resting place such as the Earth; it would be a very inefficient process, although it is true that this is how terrestial seeds scatter with some falling on stony ground. Secondly, the theory (from Hoyle) is complicated in suggesting a silicon computer arrangement in the background. And thirdly, once one has established the plausibility of a causal higher cosmic intelligence, it would seem simpler for 'seeds' to be produced by a more direct process tailored to environments.

Thus, although I agree with Hoyle that one must, somewhere along the line, introduce higher intelligence ('God' for simplicity), I have (Chapter 28) suggested that such higher intelligence is the Void of all space-time and can create simply by moulding space into forms. It is the concept that, given the Void, the subtantial universe including life can therefore be directly 'thought up' (as LOGOS). This proposal would not have been credible without Einstein's matter-tensor which shows that substance can be equated to formed space-time. But if this is so, then God can think up any constructions simply by imagining them.

So I do not think the Panspermia theory is needed, nor its silicon chip back-up; rather the DNA seeds come from 'nowhere' and are the manifestation of purposeful cosmic thought. When one is faced by supernatural intelligence, perhaps one should listen to Einstein:

An intelligence of such superiority that all the acting and thinking of human beings is an utterly insignificant reflection.

It appears that 'THE STUFF OF THE WORLD IS MIND-STUFF' , and that this applies both to the Void (God) and its shaping (LOGOS). We are virtually back to something like Genesis.

# EPILOGUE
## SCIENCE AND RELIGION

This Epilogue is concerned with answering two questions. These are: Why does religion stand at such a low level? What is the true assessment as between religion and science at the present time?

## *The factual status of religion*

The long-standing hostility between science and religion, which began with the threat of torture to Galileo by the Inquisition and which was maintained between the seventeenth and nineteenth centuries, is now largely resolved. The Church could agree that the Earth moves around the Sun although Einstein's Relativity might not invalidate the contra-view. So we are into a sort of interregnum in which neither science nor religion is prepared to be dogmatic. We shall consider whether a bridge is to be built between science and religion based on this strange new concept of supernatural science, evidenced by supernatural scientific facts.

## *Science and natural theology — religious science*

Science has always contained an element of religion because of the relationship of science to medicine. The doctors have always known that the structure of the human body and its capacity to grow and heal itself contains a strong element of the miraculous. So throughout history there has been a modicum of Natural Theology in science coming from the medical end, as is so clearly expressed by Sir Charles Sherrington in his book *Man on his Nature*. Historically this establishes a scientific belief in religion as shown by the line *A* in Fig. E.1.

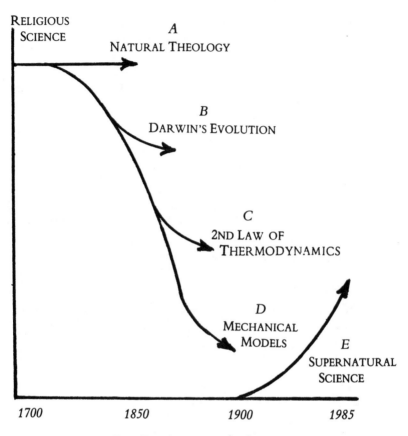

Fig. E1. Science and religion.

## Darwin's theory of the Origin of Species

The publication of *The Origin of Species* by Charles Darwin in 1859 caused a drop in the religious temperature of science (see *B* of Fig. E.1.), since Darwin introduced chance as a major factor in the development of organic life, including man. This reduced the need for God as creator, and has had a powerful effect on the thinking of scientists so that even a modern molecular biologist such as Jacques Monod could state in his book *Chance and Necessity* in 1970 (and I requote):

Man at last knows that he is alone in the unfeeling immensity of the universe, out of which he emerged only by chance.

The effect of Darwinism is to suggest that 'man comes from monkeys', so displacing a direct creative relationship between man and God.

## The Second Law of Thermodynamics

Almost contemporary with Darwin, Clausius and others were working out the laws of thermodynamics, of which the most important is the Second Law which implies that the universe is running down and will come to a full stop. This statement further lowered the religious temperature (see *C* of Fig. E.1), since one could hardly admire a creator that had created a mortal universe. We note the strong connection between Darwin's introduction of chance into biology and the Second Law, which is concerned with chance collisions in gaseous systems. The common anti-religious sentiment was 'the rule of chance'.

## Mechanical models in science

When human beings are presented with a complexity, they turn away from words and try to make a model of the situation. The sort of model which we prefer to make is 'clock-work' in which the function of each part is causally related to every other part. So the spring causes the escapement to work, which in turn causes gears to rotate, which in turn governs the position of the hands on the clock face. This sort of process is called Determinism or Necessity. The greatest success was in astronomy, which so mastered clock-work theory under Kepler and Newton that the arrival of both eclipses and comets could be forecast with uncanny accuracy.

But the defect of mechanical models of reality is that they reduce everything to the same *level* of functional uniformity and eliminate from the world-view both consciousness and values. This mechanical outlook further chilled religion, which is essentially a value system (see *D* of Fig. E.1).

## The low point of religion in 1900

From the foregoing there had been a cumulative depression of religion below the level of Natural Theology because of

1. Darwinism, which relieved the creator of his responsibilities and handed them over to random chance and 'survival of the fittest';

2. the Second Law of Thermodynamics, which insists that the universe is running down to extinction under a burden of escalating entropy; and

3. Inasmuch as the universe made sense, it was a picture of mechanicalness without values.

The overall effect on society was not only that religion was ceasing to be a significant influence, but that all the important value systems which link with religion were also deteriorating, particularly art and morality. It is not without significance that the fifty years following 1900 witnessed the greatest wars of history and the development of atomic weapons. All this was due to worshipping at the feet of Chance and Necessity.

## The coming of supernatural science — 'God exists'

The idols of Chance and Necessity are still dominant in science, at least in pedestrian science or 'scientism'. The realisation of the *fact* of supernatural science will be a shock to the scientific world, for it means that GOD EXISTS.

In considering this let us narrow down our argument for supernatural science to a single fact, which is that the specificities (improbabilities) now emerging from molecular biology involve numbers such as $10^{78,000}$ (Chapter 14), and these never could have evolved by Darwin's theory of natural selection. Let the scientists simply gaze at this one supernatural *fact* (see *E* of Fig. E.1).

## *A new model of religion — cosmic cybernetics*

If we wish to incorporate supernatural science into a religious model, we would need a model which is capable of embracing different levels as distinct from the same-level mechanical models of nineteenth-century science. Fortunately we have such models in the control systems of cybernetics. In cybernetics the models are essentially three-level as to

> Programmer
> Programme
> Data processing (print-out).

In such a triple system, which will be familiar to all who have to deal with computers, the three aspects are quite distinct, and there is a master-slave relationship between each step. So the programmer is master of the programme and the programme is master of the data processing.

## *God as programmer*

From the cybernetic analysis we need a cosmic Programmer function, and since this is the highest level of a cybernetic system, then it appears justified to equate this to God. The chief significance of a programmer is that he is *free* in at least two respects:

To programme or not to programme
To programme selectively according to conscious choice.

This freedom can be described as pure intelligence. If we wish to know 'where' God, is, I have suggested elsewhere that this is the Void — omniscient, omnipresent, invisible mind-stuff. The analogy with ourselves would be a state of lucid consciousness and capacity for thought.

## 'Logos' as programme

God as programmer writes the programme and this is LOGOS (the Word). LOGOS is the interface between God and nature. Thus LOGOS is the programme or specification for all manifestations from the cosmic constants to the DNA. It is the interface of specified specificity. As to just 'where' LOGOS might be, we look for a place from which LOGOS can irradiate nature, and this would seem to be stars in general and (in our own case) the Sun.

## Nature as data processing or print-out

Finally we look for the outputs of the cosmic computer system as physical substance and order, and this is nature. Nature is the print-out. A major place where this operates is into our biosphere.

## Can science now agree with religion?

We have seen that up until 1900 science dealt heavy blows to religion based on 'proofs' related to chance and necessity and attributable to Darwinism, the Second Law of Thermodynamics and single-level mechanical models of reality. All these three ideologies have been progressively demolished from 1900 onwards, particularly by Planck, Einstein and Heisenberg, and latterly by molecular biology. They are replaced by a new idea of the importance and dominance of *specificity* in the universe, with the inevitable implication that GOD EXISTS. What implications this may hold for humanity is outside the range of our present discussion.